International Series in Operations Research & Management Science

Volume 267

More information about this series at http://www.springernature.com/series/6161

Necmi K. Avkiran • Christian M. Ringle
Editors

Partial Least Squares Structural Equation Modeling

Recent Advances in Banking and Finance

Springer

Editors
Necmi K. Avkiran
School of Business
University of Queensland
St Lucia, QLD, Australia

Christian M. Ringle
Institute of HRM
Hamburg University of Technology
Hamburg, Germany

The University of Newcastle
Faculty of Business and Law Callaghan
Callaghan, NSW, Australia

ISSN 0884-8289 ISSN 2214-7934 (electronic)
International Series in Operations Research & Management Science
ISBN 978-3-319-71690-9 ISBN 978-3-319-71691-6 (eBook)
https://doi.org/10.1007/978-3-319-71691-6

Library of Congress Control Number: 2018931193

Printed on acid-free paper

This Springer imprint is published by Springer Nature
The registered company is Springer International Publishing AG
The registered company address is: Gewerbestrasse 11, 6330 Cham, Switzerland

Preface

We conceived this book because of the absence of *partial least squares structural equation modelling* (PLS-SEM) in banking and finance disciplines. Yet, the PLS-SEM method has been broadly accepted and used in disciplines such as accounting, health care, hospitality management, management information systems, marketing, operations management, strategic management, supply chain management, and tourism. Besides, the method enjoys an increasing application in a wide range of additional disciplines such as economics, engineering, environmental sciences, medicine, political sciences, and psychology (Richter et al. 2016). Against this background, we also expect an adoption of PLS-SEM in the banking and finance disciplines.

As a causal-predictive method, PLS-SEM has a wide spectrum of practical applications to managerial challenges. Unfortunately, secondary data frequently found in business databases are unlikely to satisfy such constraints as homogeneity in the population, and measurement errors being uncorrelated. With the ever-increasing availability of secondary data, PLS-SEM's soft modelling approach fits exploratory research, where theory has *not* been fully developed. Using the PLS-SEM approach is recommended when (a) the objective is explaining and predicting target constructs and/or detecting important driver constructs, (b) the structural model has formatively measured constructs, (c) the model is complex (with many constructs and indicators), (d) the researcher is working with a small sample size (due to a small population size), and (e) the researcher intends to use latent variable scores in follow-up studies (Sarstedt et al. 2017). PLS-SEM is relatively robust with *non-normal* data. However, researchers should not use the latter characteristic and/or small sample sizes as the sole argument for selecting PLS-SEM but focus on the goal of their empirical analysis (Rigdon 2016). This is important to proactively attend to potential criticism that has been put forward with regard to PLS-SEM (for further details on this debate, see Sarstedt et al. 2016)

The applications in this handbook further pioneer PLS-SEM adoptions in the banking and finance disciplines. New PLS-SEM developments will further expand the method's usefulness to banking and finance studies. These advances primarily address both the method's explanatory and predictive capabilities. Examples of recent enhancements include methods for uncovering unobserved heterogeneity, different multi-group analysis approaches, testing measurement invariance of composites, overall goodness-of-fit measures, and novel approaches of prediction-oriented results evaluations (Richter et al. 2016; Sarstedt et al. 2017). We also expect that the PLS-SEM method will experience extensions in the direction of longitudinal data analysis and multilevel modelling, which will become particularly beneficial for the banking and finance disciplines given the characteristics of data usually used in such studies.

As a final comment, we would like to gratefully acknowledge the reviewers who contributed to this book: Jac Birt, Allan Hodgson, Rand Low, Lin Mi, and David Smith.

St Lucia, QLD, Australia — Necmi K. Avkiran
Hamburg, Germany — Christian M. Ringle
Callaghan, NSW, Australia

References

Richter, N. F., Carrión, G. C., Roldán, J. L., & Ringle, C. M. (2016). European management research using partial least squares structural equation modeling (PLS-SEM): Editorial. *European Management Journal, 34*(6), 589–597.

Rigdon, E. E. (2016). Choosing PLS path modeling as analytical method in European management research: A realist perspective. *European Management Journal, 34*(6), 598–605.

Sarstedt, M., Hair, J. F., Ringle, C. M., Thiele, K. O., & Gudergan S. P. (2016). Estimation issues with PLS and CBSEM: Where the bias lies! *Journal of Business Research, 69*(10), 3998–4010.

Sarstedt, M., Ringle C. M., & Hair, J. F. (2017). Partial least squares structural equation modeling. In C. Homburg, M. Klarmann, & A. Vomberg (Eds.), *Handbook of market research*. Heidelberg: Springer.

Contents

Contributors

Necmi K. Avkiran School of Business, University of Queensland, St Lucia, QLD, Australia

Charmele Ayadurai School of Business and Law, University of Salford, Manchester, UK

Rasol Eskandari School of Business and Law, University of Salford, Manchester, UK

Anne-Kathrin Hegner-Kakar Hamburg University of Technology (TUHH), Hamburg, Germany

Fashion and Lifestyle, GfK SE, Nürnberg, Germany

Ralf Jasny Faculty 3: Business and Law, Frankfurt University of Applied Sciences, Frankfurt, Germany

Hengky Latan Department of Accounting, STIE Bank BPD Jateng, Semarang, Indonesia

Department of Accounting, Petra Christian University, Surabaya, Indonesia

Gilbert V. Nartea Department of Economics and Finance College of Business and Law, University of Canterbury, Christchurch, New Zealand

Christian Nitzl University of the German Armed Forces Munich, Neubiberg, Germany

Lăcrămioara Radomir Department of Marketing, Faculty of Economics and Business Administration, Babeş-Bolyai University, Cluj-Napoca, Romania

Nur Ainna Ramli Faculty of Economics and Muamalat, University Sains Islam Malaysia, Bandar Baru Nilai, Negeri Sembilan, Malaysia

Nicole F. Richter University of Southern Denmark, Sønderborg, Denmark

Sebastian Rick Governance & Assurance Services, KPMG AG Wirtschaftsprüfungsgesellschaft, The SQUAIRE, Frankfurt, Germany

Christian M. Ringle Institute of HRM, Hamburg University of Technology, Hamburg, Germany

The University of Newcastle, Faculty of Business and Law Callaghan, Callaghan, NSW, Australia

Alan Wilson Department of Marketing, Strathclyde Business School, University of Strathclyde, Glasgow, UK

Chapter 1
Rise of the Partial Least Squares Structural Equation Modeling: An Application in Banking

Necmi K. Avkiran

Abstract Researchers across a wide range of disciplines exploited the capabilities of partial least squares structural equation modeling (PLS-SEM). The rise in popularity of PLS-SEM is particularly noticeable 2013 onwards. The banking and finance discipline, however, hardly exploits the advantages of the PLS-SEM approach. PLS-SEM can be used for prediction and exploration in complex models with relaxed expectations on data. PLS-SEM is useful in identifying relationships between constructs. If the primary objective is theory development, PLS-SEM is appropriate.

Keywords Banking · Managerial competency · PLS-SEM · CB-SEM · IPMA · GSCA

1.1 Introduction

Researchers across a wide range of disciplines exploited the capabilities of partial least squares structural equation modeling (PLS-SEM) in their studies. These disciplines include, for example, accounting (Lee et al. 2011; Nitzl 2016), family business (Sarstedt et al. 2014), health care (Avkiran 2017), international business (Richter et al. 2016b), management information systems (Hair et al. 2017a; Ringle et al. 2012), marketing (Hair et al. 2012c), operations management (Peng and Lai 2012), psychology (Willaby et al. 2015), strategic management (Hair et al. 2012b), supply chain management (Kaufmann and Gaeckler 2015), and tourism (do Valle and Assaker 2016). Figure 1.1 shows the rise of PLS-SEM (e.g., Wold 1982; Lohmöller 1989), as a variance-based structural equation modeling method (SEM), in literature alongside the traditionally methodological sibling, the

N.K. Avkiran (✉)
School of Business, University of Queensland, St Lucia, QLD, Australia
e-mail: n.avkiran@business.uq.edu.au

N.K. Avkiran, C.M. Ringle (eds.), *Partial Least Squares Structural Equation Modeling*, International Series in Operations Research & Management Science 267, https://doi.org/10.1007/978-3-319-71691-6_1

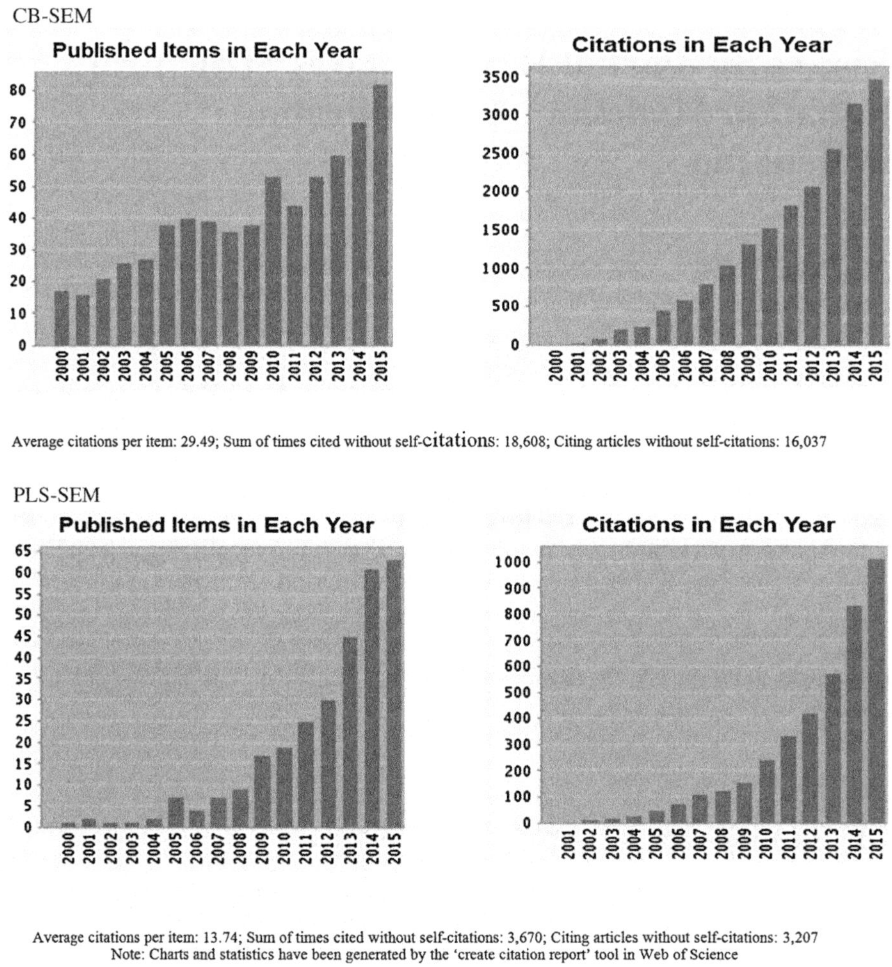

Fig. 1.1 CB-SEM versus PLS-SEM literature, 2000–2015

covariance-based structural equation modeling (CB-SEM), e.g., Jöreskog (1979). The rise in popularity of PLS-SEM is particularly noticeable 2013 onwards.

The banking and finance discipline, however, hardly exploits the advantages of the PLS-SEM approach, with few exceptions, e.g., Chu et al. (2012), Teo et al. (2015), and Kumar and Waheed (2015). A search in the reputable *Journal of Banking and Finance* reveals no PLS-SEM articles. For this reason, this chapter aims at introducing the PLS-SEM method to banking and finance. To do so, Sect. 1.2 briefly explains the PLS-SEM method and provides some advice when to choose and use this SEM method instead of its CB-SEM sibling. Section 1.3 outlines the steps to be followed in PLS-SEM analysis. Section 1.4 discusses the theoretical model behind the illustrative banking example, while Sect. 1.5 reports the results of the PLS-SEM analysis including robustness testing. Section 1.6 offers some concluding remarks.

1.2 On the Choice and Use of Partial Least Squares Structural Equation Modeling

PLS-SEM is a non-parametric, multivariate approach based on iterative OLS regression to estimate models with latent variables and their directed relationships (Wold 1982; Lohmöller 1989). Latent constructs cannot be directly observed but can be measured indirectly through several indicators (e.g., quality measured by several observed indicators based on responses from a questionnaire; for a detailed example, see Hair et al. 2017b). PLS-SEM models consist of two main components, namely, the structural model (or inner model) and the measurement models (or outer models).

More specifically, PLS-SEM estimates composite indicator models as proxies of the latent variables (Sarstedt et al. 2016). A group of indicators (manifest variables) associated with a latent construct is referred to as a *block*, and an indicator can only be associated with one construct. PLS-SEM, in its standard form, requires recursive models (i.e., there are no circular relationships or loops and the model is a predictive chain; Tenenhaus et al. 2005; Hair et al. 2017b). The primary goal of PLS-SEM is to estimate the latent variable scores that maximize the explained variance of the endogenous (i.e., dependent) latent construct(s) in the path model. These results are also used for predictive purposes (Evermann and Tate 2016; Shmueli 2016). An introduction to PLS-SEM can be found, for example, in Chin (2010), Hair et al. (2017b, 2018), Garson (2016), Sarstedt et al. (2018), and Tenenhaus et al. (2005), whereas Lohmöller (1989) and Monecke and Leisch (2012) provide a step-by-step explanation of the mathematics behind its algorithm.

As a predictive method, PLS-SEM has a wide spectrum of practical applications to managerial challenges, in particular, where human interaction is found. For example, the illustrative example used in this chapter explains managerial competence (a latent construct) by observing other latent constructs such as interpersonal skills, entrepreneurial style, and emotional maturity and experience. Such constructs are often measured indirectly through composite indicators based on metric or quasi-metric data (e.g., interval scales).

Figure 1.2 is an illustrative depiction of PLS-SEM modeling. Circles represent the latent variables or constructs that comprise the structural model; left-hand rectangles (X1–X5) house the formative indicators (composite indicators) theorized as underlying sources of the two exogenous latent constructs; right-hand rectangles (X6–X8) house the reflective indicators theorized as the consequences of the endogenous or target latent construct. W1–W8 are the outer relationships. These are often called outer weights or regression weights when the relationship is from the indicators to the construct (see constructs Y1 and Y2 and their outer weights W1–W5).

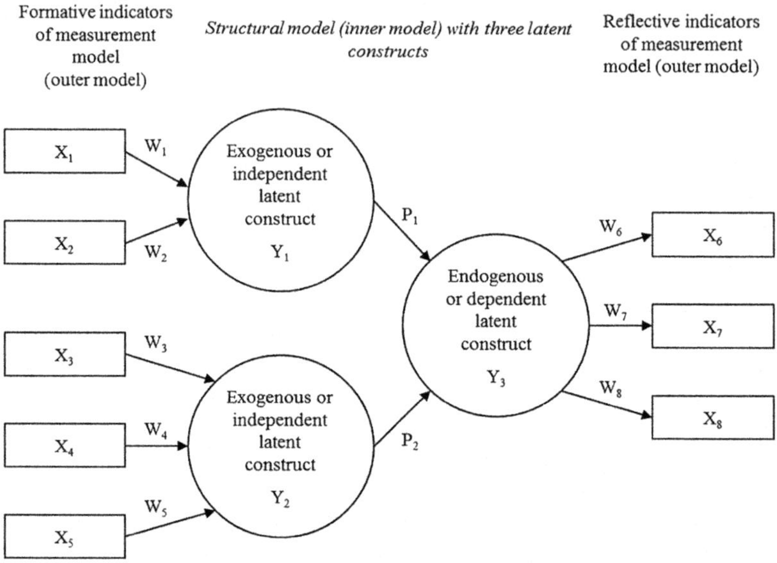

Fig. 1.2 Illustrative representation of a predictive PLS-SEM model

Otherwise, when the relationship is from the construct to the indicators, they are called outer loadings or correlation weights (see construct Y3 and the outer loadings W6–W8). P1 and P2 are the path coefficients for Y1 and Y2 (exogenous latent constructs) explaining Y3 (endogenous latent construct). The number of indicators represented in Fig. 1.2 is illustrative only.

Formative indicators are considered to represent sources that form associated exogenous latent constructs, whereby one distinguishes causal-indicator models and composite indicator models (Bollen and Diamantopoulos 2017). PLS-SEM estimates composites as proxies of (formative) composite indicator models. To establish composites, the overlap among indicators is minimized because they are considered to be complementary. The exogenous latent constructs illustrated in Fig. 1.2 become the dependent variables in multiple regression where the associated composite indicators are the independent variables (i.e., path relationships from the indicators to the constructs).

Reflective constructs, in effect, are consequences or manifestations of the underlying target latent construct (i.e., path relationships from the construct to the indicators). Because of substantial overlap among reflective indicators they are treated as interchangeable (i.e., they are expected to be highly correlated). The PLS-SEM method estimates composites as proxies of the (reflective) effect indicator models. The endogenous latent construct becomes the independent variable in single regression runs where the reflective indicators individually become the dependent variables in each run.

No matter if the PLS-SEM composites are used to represent (formative) composite indicator models or (reflective) effect indicator models, the methods either uses correlation weights or regression weights to determine the scores that serve as proxies for the theoretically/conceptually established constructs. Usually, correlation weights are employed to obtain proxies of (reflective) effect indicator constructs while regression weights are used to obtain proxies of (formative) composite indicator models. However, Becker et al. (2013) provide more details on the differences between correlation weights and regression weights when estimating composites via PLS-SEM.

Against this background, the question arises when to use PLS-SEM for the estimation of model with latent variables and their relationships, or CB-SEM. According to Lohmöller (1989), despite the similarity of the path diagram, CB-SEM and PLS-SEM basically differ in their analysis by considering what kind of information is relevant for the model: (a) CB-SEM is a covariance structure model, PLS-SEM is a data structure model; (b) CB-SEM is a factor approach to latent variables, PLS is a composite (or component) approach; (c) CB-SEM models the relationships between the variables (structural modeling), whereas PLS-SEM additionally models the relationships between the cases, thereby reconstructing/predicting the data matrix (predictive modeling). More specifically, CB-SEM is able to model measurement error structures via a factor analytic approach but at the cost of covariances among the observed variables conforming to overlapping proportionality constraints (i.e., measurement errors are assumed to be uncorrelated; Jöreskog 1979). CB-SEM assumes homogeneity in the observed population (Wu et al. 2012). Such constraints are unlikely to hold unless latent constructs are based on highly developed theory and the measurement instrument is refined through multiple stages. Thus, secondary data frequently found in business databases are unlikely to satisfy such constraints. In such cases, CB-SEM that relies on *factors* would be inappropriate, and PLS-SEM that relies on *weighted composites* would be more appropriate because of its less restrictive assumptions. Furthermore, using formative indicators is problematic in CB-SEM because it gives rise to identification problems and reduces the ability of CB-SEM to reliably capture measurement error (Petter et al. 2007). Hence, CB-SEM is more appropriate when the following conditions are satisfied: (a) main goal is theory confirmation or comparing alternative theories; (b) structural model has non-recursive relationships (i.e., circular relationships are allowed); (c) global goodness-of-fit is needed (e.g., for model comparison); (d) error terms require additional specification such as measurement of covariation. Lei and Wu (2007) provide a highly readable introduction to CB-SEM.

When one assumes a factor model as statistically estimated by the CB-SEM approach, PLS-SEM has been criticized for giving biased parameter estimates because it does not explicitly model measurement error (Gefen et al. 2011; those interested in further critique/rebuttal of PLS-SEM are encouraged to read Henseler et al. 2014). However, the bias occurs when one considers that factor models represent the optimum proxy of a latent variable. Otherwise, if one assumes composite models as suitable proxy of latent variables, PLS-SEM does not show a bias (Sarstedt et al. 2016).

Hence, one needs to distinguish the theoretical layer from the model estimation technique (i.e., a reflective conceptualized construct is not equivalent with the factor model approach for its statistical estimation), and to distinguish between the views of nominalists, operationalists, and empirical realists when establishing and estimating a model (Rigdon et al. 2017). When considering a factor model for the statistical estimation of a proxy for a latent variable, Sohn et al. (2007) restate this potential shortcoming, as PLS-SEM parameter estimates that are based on limited information not being as efficient as those based on full information estimates found in CB-SEM. However, Chin (2010) sees this as a major drawback of CB-SEM because of the assumption of the specified model as being true. That is, as a full information approach, any model misspecification in CB-SEM can impact estimates throughout the entire analysis, and unlike PLS-SEM, the overall model fit does not differentiate between the proximity of constructs.

Against this background, attempts so far to develop goodness-of-fit indices for PLS-SEM have not been entirely successful. Henseler and Sarstedt's (2013) simulation study shows that goodness-of-fit (as per Tenenhaus et al. 2004) and the relative goodness-of-fit (as per Vinzi et al. 2010) indices are not suitable for model validation. Finally, consistent PLS (PLSc) allows to correct the so-called PLS bias, and thereby, mimics CB-SEM's factor model based results (Bentler and Huang 2014; Dijkstra 2014; Dijkstra and Henseler 2015). In this kind of situation, one may revert to fit measures such as standardized root mean square residual (SRMR), normed fit index (NFI), and exact fit measures (Henseler et al. 2014).

In summary, for the method selection, Wold (2006, p. 9) notes that in large and complex models with latent variables, PLS-SEM is "virtually without competition." Researchers use this PLS-SEM characteristic to explore extensions of existing theories and to develop new and relatively complex theoretical models. According to Jöreskog and Wold (1982, p. 270) "PLS is primarily intended for causal-predictive analysis in situations of high complexity but low theoretical information." Finally, PLS-SEM is relatively robust with non-normal data (which may represent an advantage when dealing with secondary data frequently found in business databases) and small sample sizes (which may represent an advantage when samples are small due to a small population of interest for the analysis).

With the ever-increasing availability of secondary data, PLS-SEM's soft modeling approach fits exploratory research: "Soft modeling is primarily designed for research contexts that are simultaneously data-rich and theory-skeletal." (Wold 1982, p. 29). In accordance, Wold (2006) provides the following key reasons for using PLS-SEM (also see Richter et al. 2016a): (a) the PLS-SEM approach has a broad scope and flexibility of theory and practice; and (b) PLS path model improvements such as the introduction of a new latent variable, and indicator, and an inner model relation, or the omission of such an element, are easily and quickly tested for predictive relevance. Moreover, in line with, for example, Gefen et al. (2011) and Rigdon (2012, 2014), prediction-oriented analyses, complex models, and secondary data motivate the use of PLS-SEM. Additional reasons, suggested by Sarstedt et al. (2016) and Rigdon (2016), are the use of composites that represent formatively measured latent variables, the use of small sample sizes due to a small population, applying PLS-SEM latent variable scores in subsequent analyses, and endeavoring to overcome factor-based SEM's limitation by mimicking the results of factor models (i.e., by using consistent PLS approaches; Bentler and Huang 2014; Dijkstra 2014; Dijkstra and Henseler 2015). In addition, one should consider problematic arguments for and against the use of PLS-SEM that Rigdon (2016) discusses. To further substantiate the choice between both methods, Rigdon et al. (2017) draws on five perspectives to conceptually distinguish the use CB-SEM and PLS-SEM, while Lohmöller (1989) compares the statistically different estimation principles (least squares vs. maximum likelihood).

1.3 Evaluating the PLS-SEM Measurement and Structural Models

After the estimation of PLS path models, the results evaluation follows, which usually begins by focusing on the measurement models, followed by an evaluation of the structural model. For the estimation of composites to represent (reflective) effect indicator models, or (formative) composite indicator models, we outline the commonly established evaluation procedures to be followed in PLS-SEM analysis (e.g., Chin 1998, 2010; Tenenhaus et al. 2005; Haenlein and Kaplan 2004; Henseler et al. 2009; Hair et al. 2017b); we also encourage the reader to refer to Table 5 in Hair et al. (2012c) for more details on the outlined procedure. In the following, we refer to the commonly established terms of reflective and formative measurement model evaluation. Once these steps substantiate the use of the composites to represent the constructs in the model, the structural model evaluation follows.

1.3.1 Reflective Measurement Model

Each reflective indicator is related to a specific construct or latent variable by a simple regression:

$$x_h = \pi_{h0} + \pi_h \xi + \varepsilon_h$$

where $x_{h=1,\ldots,p}$is the hth regression where a reflective indicator is the dependent variable and p equals the number of reflective indicators per construct, π_{h0} is the intercept, π_h is the (single) regression parameter (outer loading) to be estimated and ξ is the latent variable. The residual variable ε_h is uncorrelated with the latent variable (also see further details on the *predictor specification* in PLS-SEM, Tenenhaus et al. 2005, p. 163).

- **Internal consistency**: The *composite reliability* criterion allows assessing the construct's internal consistency. The composite reliability should have a value of 0.7 and higher. However, values above 0.95 may indicate redundancy of items employed (Hair et al. 2017b). Composite reliability is only relevant for the reflective measurement model. Composite reliability's formula can be found, for example, in Chin (2010, p. 671), replicated below:

$$\rho_c = \frac{(\Sigma \lambda_i)^2 \operatorname{var} F}{(\Sigma \lambda_i)^2 \operatorname{var} F + \Sigma \Theta_{ii}}$$

where λ_i, F, and Θ_{ii} are the factor loading, factor variance, and unique/error variance, respectively, where i represents the indicator variable for a specific construct.

- **Indicator reliability**: Outer loadings greater than 0.7 are desirable (Hair et al. 2011). Square of this standardized outer loading represents *communality*, that is, how much of the variation in the indicator is explained by the endogenous construct, and 1 minus communality reveals the measurement error variance. If outer loadings are less than 0.4, the reflective indicator should be deleted. When an outer loading is between 0.4 and 0.7 the decision on whether to keep or delete the item depends on the (high) outer loadings of the other items and the results of criteria such as composite reliability and convergent validity (for more details, see Hair et al. 2017b).
- **Convergent validity**: Average variance extracted (AVE) greater than 0.5 is preferred; this ratio implies that greater than 50% of the variance of the reflective indicators have been accounted for by the latent variable. AVE is only relevant for the reflective measurement model. When examining reflective indicator loadings, it is desirable to see *higher loadings in a narrow range*, indicating all items are explaining the underlying latent construct (i.e., convergent validity; Chin 2010). The formula for AVE is replicated below (Chin 2010, p. 670; refer to the previous equation for variables):

$$AVE = \frac{(\Sigma \lambda_i^2) \operatorname{var} F}{(\Sigma \lambda_i^2) \operatorname{var} F + \Sigma \Theta_{ii}}$$

- **Discriminant validity**: Fornell-Larcker criterion states that the square root of AVE must be greater than the correlation of the reflective construct with all other constructs; this criterion is not applicable to formative measurement models and single-item constructs. Checking cross-loadings, all the indicators should load the highest on their associated constructs. Recently, the heterotrait-monotrait ratio of correlations (HTMT) has become the primary criterion for assessing discriminant validity since it offers superior performance compared with the Fornell-Larcker criterion and the assessment of cross-loadings (Henseler et al. 2015; also see Voorhees et al. 2016). HTMT can be computed for reflective measurement models against the threshold value of 0.90 (i.e., for discriminant validity to be established, the HTMT values should not exceed 0.90; Henseler et al. 2015).

1.3.2 Formative Measurement Model

Under the formative measurement model, it is assumed that the exogenous construct (latent variable, ξ) is defined by the formative indicators that could be multidimensional and a residual term is found in a linear function.

$$\xi = \sum_h \varpi_h x_h + \delta$$

where ϖ_h are the (multiple) regression coefficients (outer weights), the residual vector δ is uncorrelated with the formative indicators x_h where h captures the number of formative indicators per construct (Tenenhaus et al. 2005, p. 165). It is important that the researcher establishes theoretical content validity before attempting empirical analysis to ensure that the major dimensions of the construct have been covered by the indicators.

- **Convergent validity**: This analysis assesses the extent to which a measure correlates positively with other (e.g., reflective) measures of the same construct using different indicators. For this purpose, it is necessary to test whether a formative construct is strongly related with a reflective measure of the same construct (redundancy analysis). Higher path coefficients (e.g., >0.7; Hair et al. 2017b) linking the exogenous and endogenous constructs are preferred, implying adequate coverage by the formative indicators (Chin 2010). Alternatively, if a reflective construct is not available, one may link the formative measurement model with a single-item construct (i.e., the overall assessment of the construct of interest) or the content-wise closest construct available. The weakest form of convergent validity assessment analyses if estimated relationships of the formative construct with other constructs in the structural model have the expected sign (and expected rank-order of size).
- **Collinearity among indicators**: When collinearity exceeds critical levels, standard errors and thus variances are inflated. A *variance inflation factor* (VIF) is calculated as

$$VIF_i = \frac{1}{\left(1 - R_i^2\right)}$$

where R_i^2 is the proportion of variance of formative indicator i associated with other indicators in the same block (Hair et al. 2017b). Statistically, VIF is the reciprocal of *tolerance*, $\left(1 - R_i^2\right)$, where the latter is defined as the variance of a formative indicator not explained by others in the same block. A VIF of 1 means there is no correlation among the predictor variable examined and the rest of the predictors, and therefore, the variance is not inflated. For the interpretability of outer weights, the VIF must be less than 5 (Hair et al. 2011), or even better, lower than 3. If the VIF of certain indicators in the formative measurement model exceeds the critical VIF values, the researcher should consider removing the corresponding indicator, or combine the collinear indicators into a new composite indicator. If such critical VIF value is not possible (e.g., for content related or technical reasons), one can continue to use the related construct in the PLS path model but should refrain from interpreting the size of outer weights.

- **Significance and relevance of outer weights**: To assess significance of outer weights, one can use the bootstrapping routine (e.g., by building on 5000 sub-samples) in order to check whether the outer weights are significantly different from zero. Bootstrapping involves random drawing of sub-samples from the original set of data with replacement (sub-sample size equals the original sample size). 95% bias-corrected and accelerated (BCa) bootstrap confidence intervals allow assessing the significance of outer weights (i.e., with a 5% probability of error) by checking if confidence interval does not include zero. Alternatively, one can revert to *p*-values being smaller than 5% for significance testing. Indicators with significant outer weights are kept. Otherwise, it is important to distinguish between the outer weight (or regression weight) that represents an indicator's relative contribution to the construct, while the outer loading (or correlation weight) is an indicator's absolute contribution. Hence, an indicator that exhibits a non-significant outer weight can still be kept if its outer loading is greater than 0.5. However, formative indicators with non-significant outer weights and outer loadings less than 0.5 can be removed from the model for being irrelevant. Nevertheless, eliminating formative indicators should be approached with caution because formative measurement theory expects the indicators to cover the domain of a construct (i.e., formative indicators are complementary).

1.3.3 Structural Model

If the measurement (or outer) model evaluation substantiates the estimated constructs, the assessment of the structural (inner) model follows. Analysis of the structural model is an attempt to find evidence supporting the theoretical model (i.e., the theorized relationships between exogenous constructs and the endogenous construct):

$$\xi_j = \beta_{jo} + \sum_i \beta_{ji}\xi_i + v_j$$

where ξ_j is the endogenous construct and ξ_i represents the exogenous constructs, while β_{jo} is the constant term in this (multiple) regression model, β_{ij} are the regression coefficients, and v_{j} is the error term; the predictor specification condition applies (Tenenhaus et al. 2005, p. 165).

- **Collinearity**: Once again, VIF is the measure of interest. For the structural model, *inner* VIF values are examined.

- **Significance of path coefficients**: 95% bias-corrected and accelerated (BCa) bootstrap confidence intervals should be used to assess the significance of path coefficients in the structural model. Alternatively, one may revert to the bootstrap *p*-values.
- **Predictive accuracy, coefficient of determination (R^2)**: This statistic indicates to what extent the exogenous construct(s) are explaining the endogenous construct. According to Hair et al. (2011, 2017b), as a very rough guide, R^2 values of 0.25, 0.50, and 0.75 represent weak, moderate, and substantial levels. However, unless the adjusted R^2 is used (for a formal definition, see Hair et al. 2017b, p. 199), this coefficient can be upward-biased in complex models where more paths are pointing towards the endogenous construct. More importantly, coefficient of determination needs to be judged in the context of a research project's discipline to asses if the obtained R^2 value is substantial. In some disciplines, R^2 values of 0.2 are already relatively high.
- **Assessing the 'effect sizes'(f^2)**: This statistic measures the importance of the exogenous construct(s) in explaining the endogenous construct and it re-calculates R^2 by omitting one exogenous construct at a time. Again, effect size of 0.02 is small, 0.15 is moderate and 0.35 is substantial. The formula for f^2 can be found in Hair et al. (2017b, p. 201), as well as in Chin (2010, p. 675):

$$f^2 = \frac{R^2_{included} - R^2_{excluded}}{1 - R^2_{included}}$$

where $R^2_{included}$ and $R^2_{excluded}$ are the R^2 generated on the endogenous construct (dependent latent variable) when a predictor exogenous construct is used or left out in the structural equation.
- **Predictive relevance (Q^2)**: This statistic is obtained by the sample re-use technique called "Blindfolding" where *omission distance* is set between 5 and 10, where the number of observations divided by the omission distance is not an integer (Hair et al. 2012c). For example, if you select an omission distance of 7, then every seventh data point is omitted and parameters are estimated with the remaining data points. According to Hair et al. (2017b), omitted data points are considered missing values and replaced by mean values. Estimated parameters help predict the omitted data points and the difference between the actual omitted data points and predicted data points becomes the input to calculation of Q^2. *Blindfolding is applied only to endogenous constructs with reflective indicators.* If Q^2 is larger than zero, it is indicative of the path model's predictive relevance in the context of the endogenous construct and the corresponding reflective indicators. The formula follows (Chin 2010, p. 680):

$$Q^2 = 1 - \frac{\sum_D E_D}{\sum_D O_D}$$

where D is the omission distance in blindfolding, E is the sum of squares of prediction error, and O is the sum of squares errors using the mean for prediction.

- **Assessing the relative impact of predictive relevance (q^2)**: q^2 effect size provides further insights to the quality of the PLS path model estimations (Hair et al. 2017b). To ascertain the impact of a given exogenous construct on the endogenous construct, the effect size is calculated by including and excluding the exogenous construct in question where the omission distance is kept identical. Following from the above analysis of predictive relevance, q^2 effect size can be calculated as below by excluding the exogenous constructs one at a time (Hair et al. 2017b, p. 207):

$$q^2 = \frac{Q^2_{included} - Q^2_{excluded}}{1 - Q^2_{included}}$$

According to Hair et al. (2013, 2017b), effect size of 0.02 is considered small, 0.15 is moderate and 0.35 is large.

1.3.4 Importance-Performance Map Analysis (IPMA)

We take the above analysis a step further by recommending to undertake an importance-performance map analysis (IPMA; Ringle and Sarstedt 2016). The main purpose of this step is to identify areas where managerial action is likely to bring the greatest improvement of a selected target construct in the PLS path model.

Before running the analysis, the researcher needs to conform to two IPMA conditions, namely, indicators must have the same direction in terms of their meaning, and outer weights must not be negative (Hair et al. 2017b). Then, the researcher must select a specific target construct of interest. For this constructs predecessors in the PLS path model (i.e., construct that directly or indirectly explain it), the IPMA pulls together their average latent variable scores (in the range 0–100) representing *performance*, and their unstandardized total effects (sum of direct and indirect effects in the structural model) on the target construct representing *importance*. Scatter plotting the importance (i.e., the x-axis) on the performance (i.e., y-axis) results in the importance-performance map. Constructs with a relatively high importance but a relatively low performance are particularly interesting for improvements that, in turn, results in an increased performance of the selected target construct. If it is possible to increase the performance of a predecessor in the PLS path model by one point (e.g., from 78 to 79), one (ceteris paribus) assumes a performance increase of the selected target construct by the size of the predecessor's unstandardized total effect. Besides running the IPMA on the construct level, it also is possible to extend it to the indicator level.

The IPMA is particularly useful to expand the analysis of PLS-SEM results (i.e., the total effects) by the information about the performance that is included in the used data set. This kind of outcome facilitates a deeper discussion of results and to obtain more specific findings. Many applications use the IPMA to extend the PLS-SEM analysis and to advance the discussion of results. Examples in business research address customer satisfaction (Martensen and Grønholdt 2003), the satisfaction of stadium visitors (Höck et al. 2010), the reputation of corporations (Hair et al. 2018), and non-for-profit organizations (Schloderer et al. 2014), and the acceptance of technology (Ringle and Sarstedt 2016).

1.4 Theoretical Model Behind the Illustrative Banking Example

Rose (1986) argues that a bank branch's performance is affected by competence of its management, as well as economic and social conditions prevailing in its catchment area. Probably one of most influential controllable variables in bank branch performance is the *managerial competence* of the branch manager, where the manager is seen as a team leader. In the role of the team leader, the branch manager's competence will influence the full range of branch activities. The study of managerial competence is particularly appropriate if the executive managers of a bank are interested in prospective information that can be used in short- to medium-term strategic planning. Thus, an examination of managerial competence can provide insights into why a branch is contributing only a certain amount to the organization and provide direction for remedial action. On the other hand, accounting indicators alone cannot be confidently used for planning purposes due to the retrospective nature of such information.

Managerial competency can be defined as "...an underlying characteristic of a manager that results in effective and/or superior performance in a managerial job" (Boyatzis 1982, p. 21). That is, managerial competency can be treated as a latent variable or construct. Avkiran (1995) uses Chataway's (1982) findings from a naturalistic inquiry that identifies needed competencies, and immediate subordinates of the branch manager become the appraisers. In Avkiran (1995), based on Australian bank branches, questionnaire recipients were asked "To what extent does your branch manager possess the competencies listed?" A five point scale (i.e., "*Not at all (0)*" to "*To a great extent (4)*") was provided for rating. Initially, 48 items on managerial competency of the branch manager were conceptualized across five dimensions. The 58 variables were then factor analyzed by Avkiran (1995) through principal axis factoring. The factor matrix was rotated orthogonally using the Varimax method. Seven extracted factors explained 62.3% of variance. Variables loading less than 0.5 were omitted in the final analysis; this meant that only 45 of the variables across three factors were retained. Consistent with the rotated factor matrix, operational definitions of the scale factors were re-stated as:

F1: *Interpersonal Skills (IPS)*: Ability to respond to staff's needs positively, fostering a non-discriminatory work environment where staff can develop to their full personal potentials, and delegating authority.
F2: *Entrepreneurial Style (ES)*: Particular disposition towards performing a range of tasks including delivery of customer service, identifying market opportunities, practicing proactive decision-making, setting achievable goals, motivating staff to work as a team, and dealing with change.
F3: *Emotional Maturity and Experience (EME)*: Ability to focus on central issues under pressure while maintaining a sense of humor, demonstrating initiative, perseverance, and knowledge of banking.

The above three factors become the exogenous constructs in PLS-SEM predictive modeling explaining the endogenous construct *managerial competency* (MC). For the purpose of illustrating PLS-SEM in banking, only the five top loading variables on each of the three factors are used as formative indicators of the three exogenous constructs. The five reflective indicators for the endogenous construct MC are selected from bank branch performance measures (i.e., reflective indicators are the consequences or manifestations of the underlying target latent construct). Table 1.1 shows the list of illustrative formative and reflective indicators; note that Table 1.1 also provides an explanation of indicators while we use their abbreviations in the remainder of this chapter.

As a result of the above theoretical discussion, three hypotheses emerge:

H_1: Managerial competency (MC) is positively and significantly explained by *interpersonal skills* (IPS).
H_2: Managerial competency (MC) is positively and significantly explained by *entrepreneurial style* (ES).
H_3: Managerial competency (MC) is positively and significantly explained by *emotional maturity and experience* (EME).

PLS-SEM analysis is used with the overall objective of rejecting null hypotheses regarding path relationships between constructs and accepting the three alternative hypotheses outlined above.

1.5 PLS-SEM Analysis

Preparation of a data file for SmartPLS requires use of columns with text headings in a spreadsheet saved as a *csv* file; the first column needs to consist of numbers rather than text (e.g., bank names cannot be used but they can be represented by numbers). Data of 101 bank branches represents an adequate sample for an initial PLS-SEM analysis to explain the managerial competence in the banking industry. Subsequent studies, however, should aim at generating more response to improve representativeness of the data and to facilitate additional analyses for specific groups of data.

Table 1.1 Illustrative formative and reflective indicators

Sources of managerial competence (i.e., the key competencies from the survey questionnaire become the formative indicators)	*Consequences* of managerial competence (i.e., the reflective indicators from the endogenous construct)[d]
Interpersonal Skills—IPS (exogenous construct)[a]	1. Number of new deposit accounts [NEWDEPO#]
1. Generates ideas for use at appropriate time (IPS_1)	
2. Conveys staff's concerns to higher management (IPS_2)	
3. Can communicate orally for two-way understanding (IPS_3)	
4. Exercises common sense in decision-making and problem-solving (IPS_4)	2. Average lending balances outstanding [LENDBAL$]
5. Is sensitive in implementing executive decisions (IPS_5)	
Entrepreneurial Style—ES (exogenous construct)[b]	
1. Identifies deposit gathering opportunities (ES_1)	
2. Develops strategies to maximize deposit gathering (ES_2)	3. Number of new home loans [HLOAN#]
3. Establishes action priorities for achievable goals (ES_3)	
4. Encourages customers to provide feedback on quality of service (ES_4)	
5. Matches management style to changing situations (ES_5)	
Emotional Maturity and Experience - EME (exogenous construct)[c]	4. Fee income [FEEINC$]
1. Has knowledge of banking culture: its norms, values, attitudes, customs and language (EME_1)	
2. Has knowledge of banking procedures: work flows, systems processing, and technicalities (EME_2)	
3. Perseveres with an issue to its conclusion (EME_3)	
4. Remains stable in pressure situations (EME_4)	5. Number of new insurance policies originated [INSURE#]
5. Focuses on central issues (EME_5)	

[a]See Table 5.9 and Appendix G in Avkiran (1995). The top loading five variables are 36, 30, 35, 37, 20
[b]12, 13, 8, 3, 14
[c]44, 45, 7, 43, 39
[d]See Table 6.5 in Avkiran (1995)

Technically, with regards to the PLS path model and the statistical method employed, the data of 101 bank branches is adequate in relation to the maximum number of indicators associated with a construct and the minimum sample size suggested by Barclay et al. (1995), i.e., 5 × 10 = 50. Referring to Exhibit 1.7 in Hair et al. (2017b), the alternative minimum sample size emerges as 20 assuming a statistical power of 80%, significance level of 5% and a minimum R^2 of 50%. Thus, the actual sample size, while small by CB-SEM standards, passes both tests on minimum sample size for PLS-SEM. Summary statistics reported in Table 1.2 indicate substantially non-normal data.

Table 1.2 Descriptive statistics on indicators used in PLS-SEM (N = 101)

	Mean	Median	Std. Dev.	CV[a]	Minimum	Maximum	Skewness	Excess kurtosis
Reflective indicators								
1. NEWDEPO#	1763	1537	916	0.52	421	5032	1.51	2.39
2. LENDBAL$	72,165,556	64,179,375	39,800,183	0.55	15,940,000	200,722,500	1.03	0.76
3. HLOAN#	156.61	133.50	85.21	0.54	44.00	475.00	1.28	1.67
4. FEEINC$	67,663	58,579	41,149	0.61	13,175	209,071	1.54	2.47
5. INSURE#	49.62	44.00	30.61	0.62	12.00	154.00	1.45	1.98
Formative indicators								
1. IPS_1	2.67	3.00	1.06	0.40	0.00	4.00	−0.84	0.47
2. IPS_2	2.57	3.00	1.21	0.47	0.00	4.00	−0.73	−0.20
3. IPS_3	3.03	3.00	0.99	0.33	0.00	4.00	−1.19	1.29
4. IPS_4	3.09	3.00	1.02	0.33	0.00	4.00	−1.39	1.84
5. IPS_5	2.73	3.00	1.20	0.44	0.00	4.00	−0.88	−0.06
6. ES_1	2.57	3.00	1.03	0.40	0.00	4.00	−0.38	−0.36
7. ES_2	2.32	2.00	1.12	0.48	0.00	4.00	−0.41	−0.28
8. ES_3	2.57	3.00	0.98	0.38	0.00	4.00	−0.58	0.26
9. ES_4	2.19	2.00	1.11	0.51	0.00	4.00	−0.20	−0.62
10. ES_5	2.59	3.00	1.16	0.45	0.00	4.00	−0.53	−0.45
11. EME_1	3.31	3.00	0.89	0.27	0.00	4.00	−1.87	4.38
12. EME_2	2.90	3.00	1.03	0.35	0.00	4.00	−0.80	0.24
13. EME_3	3.14	4.00	1.10	0.35	0.00	4.00	−1.24	0.85
14. EME_4	2.98	3.00	1.01	0.34	0.00	4.00	−1.24	1.65
15. EME_5	2.89	3.00	0.95	0.33	0.00	4.00	−1.39	2.39

[a]Coefficient of variation (std. dev./mean)

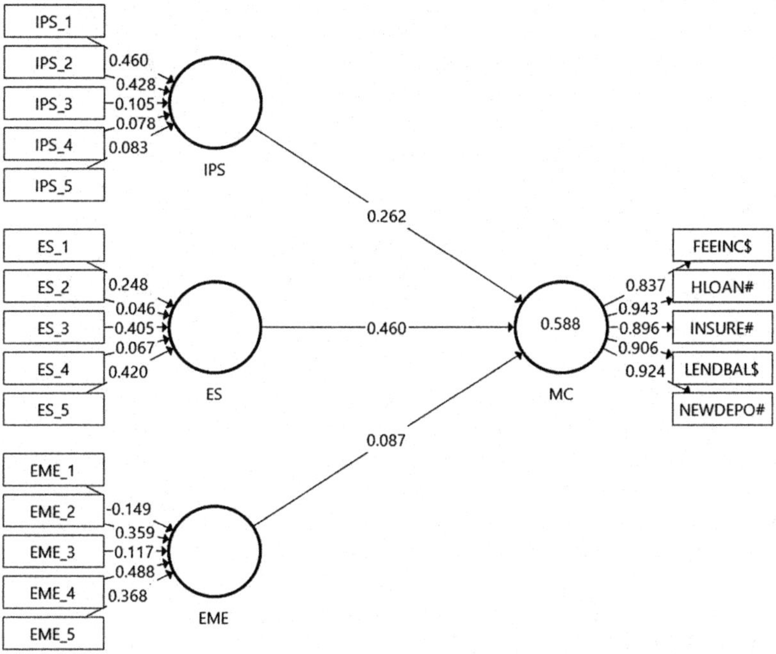

Fig. 1.3 Original PLS-SEM analysis of branch manager competency

The goal of this analysis is to explain and to predict the endogenous construct MC in the PLS path model through three exogenous constructs IPS, ES, and EME. We use the available data and the software SmartPLS 3 (Ringle et al. 2015) for conducting the PLS-SEM analysis presented in this chapter.[1] Figure 1.3 shows the PLS path model and the and estimated path coefficients.

Next, we follow the evaluation steps already detailed in Sect. 1.3. It is important to check that the algorithm converges before reaching the maximum number of iterations set in the software. Table 1.3 presents the results of the reflective and formative measurement evaluation as well as the assessment of the structural model; in addition, Table 1.4 shows the bootstrapping results of the measurement models and the structural model to determine the significance of the estimated path coefficients. With few exceptions, the results meet all relevant criteria.

Removal of the nonsignificant exogenous construct EME to some extent raises the path coefficients for the remaining two exogenous constructs with minimal impact on R^2 (Fig. 1.4). The exogenous construct of *entrepreneurial style* maintains its dominance in explaining managerial competence. In terms of the three hypotheses posed at the end of Sect. 1.4, we only reject H_3.

---

[1]A video recording of PLS-SEM analysis using SmartPLS and *R* code can be viewed at https://youtu.be/SzQ_LJWnqgQ; this recording is based on the article that can be downloaded from http://papers.ssrn.com/sol3/papers.cfm?abstract_id=2634184.

Table 1.3 Results of the original PLS-SEM analysis

Reflective measurement model evaluation	**Criterion and critical value**	**Results**	**Criterion met?**
Internal consistency	Cronbach's alpha > 0.7, <0.90 Composite reliability > 0.7, <0.90	Cronbach's alpha of 0.942 for MC Composite reliability of 0.956 MC	Yes, but a bit high Yes, but a bit high These results suggest there may be slight redundancy among the reflective indicators
Indicator reliability	Outer loadings >0.7	Outer loadings range from 0.837 to 0.943	Yes
Convergent validity	AVE >0.5	AVE = 0.813	Yes
Discriminant validity	HTMT <0.90	Not available	Not applicable (the model under study is not comprised of only reflective constructs)
Formative measurement model evaluation	**Criterion and critical value**	**Results**	**Criterion met?**
Convergent validity	Correlation between the formative and reflective (or single-item) measurement of the same construct >0.7	Not available	Data for a reflective (or single-item measurement) of the formative constructs is not available in this example. At least, the formative constructs have the expected signs in the structural model
Collinearity among indicators	VIF < 5	VIF values range from 1.589 to 4.218	Yes

Significance and relevance of outer weights (bootstrapping)	Significance: Outer weights do not fall into the 95% bootstrap confidence interval; relevance: outer loadings of the formative indicators should be above 0.5	The outer weights of indicators IPS_1, IPS_2, ES_1, ES_3, ES_5, EME_2, EME_4, and EME_5 are significant based on the 95% bootstrap confidence interval (5000 bootstraps, two-sided test, BCa method), *see* Table 1.4; all outer loadings of the formative indicators are above 0.5	Yes, for most indicators; the non-significant indicators have outer loadings larger than 0.5 and, thus, are kept in the model
Structural model	**Criterion and critical value**	**Results**	**Criterion met?**
Collinearity	VIF <5	VIF values are in the range 3.416–3.881	Yes
Significance and relevance of path coefficients (inner weights)	Significance: Inner weights do not fall into the 95% bootstrap confidence interval; relevance: inner weights > 0.1	The inner weights of the IPS to MC relationship (0.262) and the ES to MC relationship (0.460) are significant based on the 95% bootstrap confidence interval (5000 bootstraps, two-sided test, BCa method), *see* Table 1.4; the EME to MC relationship (0.087) is not significant	Yes, except for EME which is not significant and relevant for the explanation of MC
Coefficient of determination	$R^2 > 0.25$	0.588 (adjusted 0.575) for managerial competence	Yes
Effect size f^2	f^2 values larger than 0.02, 0.15 and 0.35 represent small, moderate and substantial effect sizes.	IPS: 0.043 (weak)	Yes, except for EME which does not play a relevant role in explaining MC
		ES: 0.139 (weak to moderate)	
		EME: 0.005 (not relevant)	
Predictive relevance Q^2	Q^2 value > 0	0.460 for managerial competence	Yes
Relative impact of predictive relevance q^2	q^2 values larger than 0.02, 0.15 and 0.35 represent small, moderate and substantial predictive relevance.	IPS: 0.0259 (weak)	Yes, except for EME which does not play a relevant role in predicting MC
		ES: 0.0815 (weak)	
		EME: 0.0019 (not relevant)	

IPS interpersonal skills, *ES* entrepreneurial style, *MC* managerial competence, *EME* emotional maturity and experience

Table 1.4 Bootstrapping results of path coefficients in the measurement and the structural models

	Original sample (O)	Sample mean (M)	Bias	Lower bound 2.50%	Upper bound 97.50%	Significant?
Measurement Models						
EME_1 -> EME	−0.149	−0.151	−0.002	−0.503	0.161	No
EME_2 ->EME	0.359	0.351	−0.008	0.060	0.632	Yes
EME_3 -> EME	0.117	0.113	−0.005	−0.090	0.322	No
EME_4 -> EME	0.488	0.490	0.002	0.210	0.706	Yes
EME_5 -> EME	0.368	0.375	0.007	0.089	0.654	Yes
ES_1 -> ES	0.248	0.263	0.016	0.011	0.590	Yes
ES_2 -> ES	0.046	0.038	−0.008	−0.249	0.397	No
ES_3 -> ES	0.405	0.396	−0.010	0.194	0.630	Yes
ES_4 -> ES	0.067	0.064	−0.004	−0.159	0.308	No
ES_5 -> ES	0.420	0.410	−0.010	0.139	0.689	Yes
FEEINC$ <- MC	0.190	0.191	0.001	0.164	0.207	Yes
HLOAN# <- MC	0.238	0.238	0.000	0.220	0.261	Yes
INSURE# <- MC	0.211	0.214	0.002	0.193	0.225	Yes
IPS_1 -> IPS	0.460	0.477	0.017	0.248	0.689	Yes
IPS_2 -> IPS	0.428	0.413	−0.015	0.202	0.675	Yes
IPS_3 -> IPS	0.105	0.104	0.000	−0.184	0.367	No
IPS_4 -> IPS	0.078	0.065	−0.013	−0.239	0.418	No
IPS_5 -> IPS	0.083	0.088	0.004	−0.251	0.401	No
LENDBAL$ <- MC	0.202	0.203	0.000	0.172	0.221	Yes
NEWDEPO# <- MC	0.265	0.262	−0.002	0.243	0.305	Yes
Structural model						
EME -> MC	0.087	0.099	0.012	−0.136	0.284	No
ES -> MC	0.460	0.474	0.013	0.277	0.662	Yes
IPS -> MC	0.262	0.260	−0.002	0.044	0.445	Yes

Note that the bootstrapping results have been obtained by using the following option: bias-corrected and accelerated (BCa) bootstrap approach, 5000 subsamples, no sign changes, two-sided test, 95% (bias-corrected) confidence interval

In order to further streamline the model, we address the slight redundancy observed among the reflective measurement model's indicators by removing two indicators that are highly correlated, i.e., NEWDEPO# is correlated with HLOAN# at 0.883, and LENDBAL$ is correlated with FEEINC$ at 0.869, and we delete

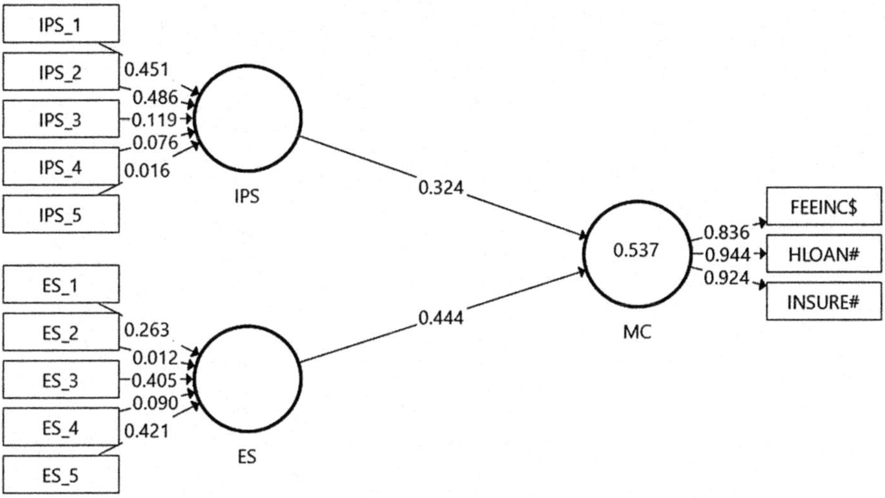

Fig. 1.4 PLS-SEM analysis of the streamlined model

NEWDEPO# and LENDBAL$. As a result, the composite reliability improves by dropping to 0.929 (below 0.95) and Cronbach alpha is now below 0.9 at 0.885. A further improvement can be observed in more balanced path coefficients (see Fig. 1.4).

1.5.1 Importance-Performance Map Analysis (IPMA)

Figure 1.5 shows the results of the construct level IPMA on the streamlined model for the endogenous construct *managerial competence*. Total effects on the horizontal axis represent *importance* and the vertical axis represents the percent performance of the two exogenous constructs in explaining the endogenous construct. While the performance of the two exogenous constructs is similar, clearly *interpersonal skills* construct is not comparably important.

In terms of raising performance, it would be better for the management to focus efforts on the *entrepreneurial style,* in the knowledge that it has a higher importance and improvements here are likely to lead to larger improvements in explaining the endogenous construct. All else the same, a one unit rise in the performance of interpersonal skills would bring about a 0.29 increase in the performance of managerial competence (see total effects). To provide a more specific guidance for management, an indicator level IPMA can also be separately undertaken (see, for example, Hair et al. 2018; Ringle and Sarstedt 2016).

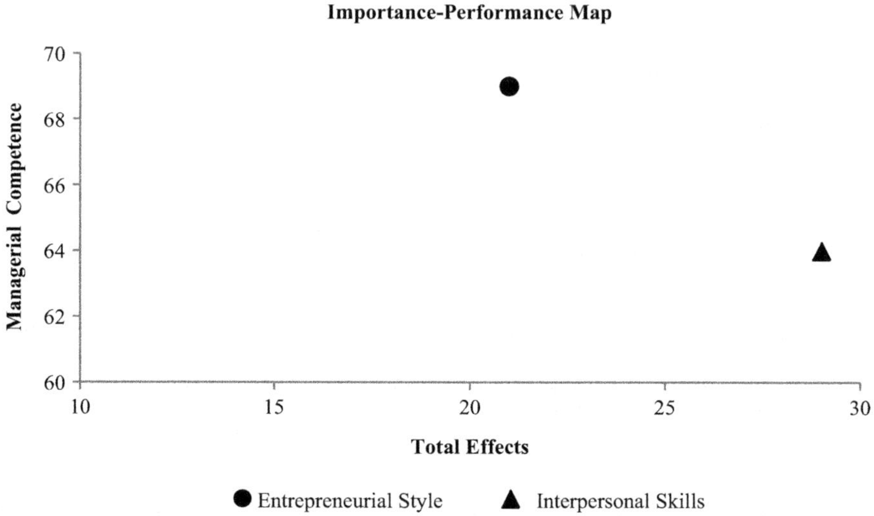

Fig. 1.5 Importance-performance map analysis

1.5.2 Robustness Testing

Generalized structured component analysis (GSCA) was introduced by Hwang and Takane (2004, 2014) as an alternative to PLS-SEM. We apply GSCA as a robustness test because it belongs to the same family of methods (Hair et al. 2017c; Schlittgen 2017). Both PLS-SEM and GSCA are variance-based methods appropriate for predictive modeling and they substitute components for factors. In terms of model specification, PLS-SEM has two equations and GSCA has one, and GSCA uses a global optimization function in parameter estimation with least squares (see Tables 1.1 in Hwang et al. 2010, and Thiele et al. 2015). We reiterate that CB-SEM is not a viable or meaningful alternative to PLS-SEM under the conditions of the current study where the sample size is small, formative indicators are present, and the study is exploratory rather than confirmatory.

GSCA maximizes the average or the sum of explained variances of linear composites, where latent variables are determined as weighted components or composites of observed variables. The technique follows a global least squares optimization criterion, which in turn, is minimized to generate the model parameter estimates. GSCA is not scale-invariant and it standardizes data. The method is supposed to retain the advantages of PLS-SEM such as less restrictions on distributional assumptions (i.e., multivariate normality of observed variables is not required for parameter estimation), unique component score estimates, and avoidance of improper solutions with small samples (Hwang and Takane 2004; Hwang et al. 2010), while additionally providing an overall measure of model fit.

We use the web based GSCA software GeSCA (http://www.sem-gesca.org/) for robustness testing of the streamlined model with three reflective indicators and ten

Table 1.5 PLS-SEM versus GSCA using the streamlined model

	PLS-SEM	GSCA
Measurement model		
Average variance extracted (AVE)	0.815	0.814
Outer loadings of reflective indicators		
FEEINC$	0.836	0.833
HLOAN#	0.944	0.952
INSURE#	0.924	0.917
Structural model		
Significant path coefficients	Both	Both
Coefficient of determination (R^2)	0.537	0.550

formative indicators across two exogenous constructs. As expected (Hair et al. 2017c), the PLS-SEM results are confirmed by GSCA (see Table 1.5). For example, AVE is almost identical; outer loadings are of similar magnitude across the three reflective indicators; both of the path coefficients are statistically significant in the structural model; and, the coefficients of determination are close to each other, with GSCA giving a slightly larger R^2.

1.6 Concluding Remarks

Our research introduces the PLS-SEM method to the banking discipline. PLS-SEM shows that close to 60% of the variation in managerial competence can be explained by a parsimonious model of ten formative indicators defining interpersonal skills and entrepreneurial style, and three reflective indicators. The path coefficients indicate that the exogenous construct *entrepreneurial style* (ES) plays a greater role in explaining managerial competence reflected in financial performance measures. These results are dependable because various consistency, reliability, validity and significance tests are satisfactory, and robustness testing using GSCA confirms the main PLS-SEM findings.

When using the SEM method, Rigdon (2016) provides a critical and philosophical examination of PLS-SEM and CB-SEM, and attempts to disassemble some of the myths perpetuated by these opposing camps (also see Rigdon et al. 2017). For example, Rigdon maintains that regardless of which SEM is used, if the sample size is small, the best approach is to collect more data. The author further points out that both SEM methods form proxies (rather than conceptual variables) out of data; for valid findings, such proxies need to be well-founded representations of conceptual variables. Statistical methods' performance drops when misspecification is present. Rigdon also states that absence of a reliable measurement of error in PLS-SEM is not a valid objection to its use because neither PLS-SEM nor CB-SEM can remove the impact of measurement error on results. Moreover, Richter et al. (2016b) examine the use of SEM in international business research and conclude that PLS-SEM is not fully utilized in the theorizing process. The authors point out that exploring is the first step in theory building where one develops hypotheses rather than test them, and PLS-SEM can be used for prediction and exploration in complex models with relaxed expectations on data. Some of the findings of the authors indicate poor compliance with basic PLS-SEM guidelines such as using a holdout sample, fully reporting the distribution of data, substantiating measurement mode, detailing contribution of indicators, identifying collinearity and reporting effect sizes. The authors acknowledge that PLS-SEM is useful in identifying relationships between constructs and explaining such relationships. They conclude their study by recommending that a study's purpose and theoretical basis should be the main selection criteria between PLS-SEM and CB-SEM. That is, if the primary objective is theory development than PLS-SEM is better, where other issues such as sample size, distributional assumptions etc. should be of secondary concern.

There are clearly differing views in literature regarding PLS-SEM and CB-SEM and the debate is not likely to be finalized soon. In conclusion, the choice between PLS-SEM and CB-SEM depends on the goal of analysis (construct explanation and prediction versus model testing and comparison), the types and numbers of latent constructs modelled (i.e., formative versus reflective, and complexity), data characteristics, and sample size. With regards to the statistical method used, it is important to note that, technically, CB-SEM estimates factor models while PLS-SEM estimates composite models. Both approaches provide different estimated proxies for the latent variables considered in the model (Sarstedt et al. 2016). Given such multiple considerations, it is best to consider CB-SEM and PLS-SEM as *complementary* rather than competitors or adversaries (Jöreskog and Wold 1982; Hair et al. 2012a).

When exploiting the advantages of PLS-SEM, researchers may consider advances on the method. These advances include the confirmatory tetrad analysis for testing the kind of measurement model and construct (Gudergan et al. 2008; Hair et al. 2018), different multigroup analysis approaches (Chin and Dibbern 2010; Hair et al. 2018; Sarstedt et al. 2011), testing measurement invariance of composites (Hair et al. 2018; Henseler et al. 2016), and methods for uncovering unobserved heterogeneity (e.g., Becker et al. 2013; Hair et al. 2016; Matthews et al. 2016; Ringle et al. 2014; Schlittgen et al. 2016). Especially methods to uncover unobserved heterogeneity should be used in any PLS-SEM analysis. Thereby, researchers can either confirm that heterogeneity does not represent a critical issue (i.e., they can conclude that the results of their sample are valid for the population).

Alternatively, if heterogeneity represents a critical issue, they can form groups of data to obtain valid group-specific results. However, not accounting for heterogeneity may represent a serious threat to the validity of results (Becker et al. 2013; Hair et al. 2018). Also, the weighted PLS algorithm (WPLS) allows to ensure validity of results (Becker and Ismail 2016). A weights vector with a specific weight for every single observation in the sample ensures repetitiveness with regards to the population. The WPLS algorithm considers these weights in all computations and thereby provides results that support representativeness and validity. Finally, the consistent PLS methods and new fit measures allow to fully mimic CB-SEM results if desired by the researcher (Dijkstra and Henseler 2015). These extensions are also useful for research in banking and finance. Future research attempts will further extend the PLS-SEM capabilities for prediction (e.g., Cepeda Carrión et al. 2016; Shmueli et al. 2016) and the use of longitudinal data. With these capabilities, PLS-SEM becomes even more relevant for research in banking and finance.

References

Avkiran, N. K. (1995). *A multivariate model of integrated branch performance and potential focusing on personal banking*. Ph.D. Thesis, Victoria University, Australia. Retrieved from http://www.users.on.net/~necmi/financesite/profile.htm.

Avkiran, N. K. (2017). An in-depth discussion and illustration of partial least squares structural equation modeling in health care. *Health Care Management Science*. https://doi.org/10.1007/s10729-017-9393-7.

Barclay, D. W., Higgins, C. A., & Thompson, R. (1995). The partial least squares approach to causal modeling: Personal computer adoption and use as illustration. *Technology Studies, 2*(2), 285–309.

Becker, J.-M., & Ismail, I. R. (2016). Accounting for sampling weights in PLS path modeling: Simulations and empirical examples. *European Management Journal, 34*(6), 606–617.

Becker, J.-M., Rai, A., Ringle, C. M., & Völckner, F. (2013). Discovering unobserved heterogeneity in structural equation models to avert validity threats. *MIS Quarterly, 37*(3), 665–694.

Bentler, P. M., & Huang, W. (2014). On components, latent variables, PLS and simple methods: Reactions to Rigdon's rethinking of PLS. *Long Range Planning, 47*(3), 138–145.

Bollen, K. A., & Diamantopoulos, A. (2017). In defense of causal-formative indicators: A minority report. *Psychological Methods, 22*(3), 581–596. https://doi.org/10.1037/met0000056.

Boyatzis, R. E. (1982). *The competent manager*. New York: Wiley.

Cepeda Carrión, G., Henseler, J., Ringle, C. M., & Roldán, J. L. (2016). Prediction-oriented modeling in business research by means of PLS path modeling. *Journal of Business Research, 69*(10), 4545–4551.

Chataway, J. G. (1982). *Bank work: A study of the everyday work activity of four suburban branch managers*. Ph.D. Thesis, Monash University.

Chin, W. W. (1998). The partial least squares approach to structural equation modeling. In G. A. Marcoulides (Ed.), *Modern methods for business research* (pp. 295–358). Mahwah: Erlbaum.

Chin, W. W. (2010). How to write up and report PLS analyses. In V. E. Vinzi, W. W. Chin, J. Henseler, & H. Wang (Eds.), *Handbook of partial least squares: Concepts, methods and applications: Vol. 2. Springer handbooks of computational statistics series* (pp. 655–690). Heidelberg: Springer.

Chin, W. W., & Dibbern, J. (2010). A permutation based procedure for multi-group PLS analysis: Results of tests of differences on simulated data and a cross cultural analysis of the sourcing of information system services between Germany and the USA. In V. Esposito Vinzi, W. W. Chin, J. Henseler, & H. Wang (Eds.), *Handbook of partial least squares: Concepts, methods and applications: Vol. 2. Springer handbooks of computational statistics series* (pp. 171–193). Heidelberg: Springer.

Chu, P. Y., Lee, G. Y., & Chao, Y. (2012). Service quality, customer satisfaction, customer trust, and loyalty in an e-banking context. *Social Behavior and Personality: An International Journal, 40*(8), 1271–1283.

Dijkstra, T. K. (2014). PLS' Janus face—Response to Professor Rigdon's 'Rethinking Partial Least Squares Modeling: In Praise of Simple Methods'. *Long Range Planning, 47*(3), 146–153.

Dijkstra, T. K., & Henseler, J. (2015). Consistent partial least squares path modeling. *MIS Quarterly, 39*(2), 297–316.

do Valle, P. O., & Assaker, G. (2016). Using partial least squares structural equation modeling in tourism research: A review of past research and recommendations for future applications. *Journal of Travel Research, 55*(6), 695–708.
Evermann, J., & Tate, M. (2016). Assessing the predictive performance of structural equation model estimators. *Journal of Business Research, 69*(10), 4565–4582.
Garson, G. D. (2016). *Partial least squares regression and structural equation models*. Asheboro: Statistical Associates.
Gefen, D., Rigdon, E. E., & Straub, D. W. (2011). Editor's comment: An update and extension to SEM guidelines for administrative and social science research. *MIS Quarterly, 35*(2), iii–xiv.
Gudergan, S. P., Ringle, C. M., Wende, S., & Will, A. (2008). Confirmatory tetrad analysis in PLS path modeling. *Journal of Business Research, 61*(12), 1238–1249.
Haenlein, M., & Kaplan, A. M. (2004). A Beginner's guide to partial least squares analysis. *Understanding Statistics, 3*(4), 283–297.
Hair, J. F., Ringle, C. M., & Sarstedt, M. (2011). PLS-SEM: Indeed a silver bullet. *Journal of Marketing Theory and Practice, 19*(2), 139–151.
Hair, J. F., Ringle, C. M., & Sarstedt, M. (2012a). Partial least squares: The better approach to structural equation modeling? *Long Range Planning, 45*(5–6), 312–319.
Hair, J. F., Sarstedt, M., Pieper, T. M., & Ringle, C. M. (2012b). The use of partial least squares structural equation modeling in strategic management research: A review of past practices and recommendations for future applications. *Long Range Planning, 45*(5–6), 320–340.
Hair, J. F., Sarstedt, M., Ringle, C. M., & Mena, J. A. (2012c). An assessment of the use of partial least squares structural equation modeling in marketing research. *Journal of the Academy of Marketing Science, 40*(3), 414–433.
Hair, J. F., Ringle, C. M., & Sarstedt, M. (2013). Partial least squares structural equation modeling: Rigorous applications, better results and higher acceptance. *Long Range Planning, 46*, 1–12.
Hair, J. F., Sarstedt, M., Matthews, L., & Ringle, C. M. (2016). Identifying and treating unobserved heterogeneity with FIMIX-PLS: Part I—method. *European Business Review, 28*(1), 63–76.
Hair, J. F., Hollingsworth, C. L., Randolph, A. B., & Chong, A. Y. L. (2017a). An updated and expanded assessment of PLS-SEM in information systems research. *Industrial Management & Data Systems, 117*(3), 442–458.
Hair, J. F., Hult, G. T. M., Ringle, C. M., & Sarstedt, M. (2017b). *A primer on partial least squares structural equation modeling (PLS-SEM)* (2nd ed.). Thousand Oaks: Sage.
Hair, J. F., Hult, G. T. M., Ringle, C. M., Sarstedt, M., & Thiele, K. O. (2017c). Mirror, mirror on the wall: A comparative evaluation of composite-based structural equation modeling methods. *Journal of the Academy of Marketing Science, 45*(5), 616–632.
Hair, J. F., Sarstedt, M., Ringle, C. M., & Gudergan, S. P. (2018). *Advanced issues in partial least squares structural equation modeling (PLS-SEM)*. Thousand Oaks: Sage.
Henseler, J., & Sarstedt, M. (2013). Goodness-of-fit indices for partial least squares path modeling. *Computational Statistics, 28*, 565–580.
Henseler, J., Ringle, C. M., & Sinkovics, R. R. (2009). The use of partial least squares path modeling in international marketing. In R. R. Sinkovics & P. N. Ghauri (Eds.), *New challenges to international marketing: Advances in international marketing* (Vol. 20, pp. 277–319). Bingley: Emerald Group.
Henseler, J., Dijkstra, T. K., Sarstedt, M., Ringle, C. M., Diamantopoulos, A., Straub, D. W., Ketchen, D. J., Hair, J. F., Hult, G. T. M., & Calantone, R. J. (2014). Common beliefs and reality about partial least squares: Comments on Rönkkö & Evermann (2013). *Organizational Research Methods, 17*(2), 182–209.
Henseler, J., Ringle, C. M., & Sarstedt, M. (2015). A new criterion for assessing discriminant validity in variance-based structural equation modeling. *Journal of the Academy of Marketing Science, 43*, 115–135.
Henseler, J., Ringle, C. M., & Sarstedt, M. (2016). Testing measurement invariance of composites using partial least squares. *International Marketing Review, 33*(3), 405–431.
Höck, C., Ringle, C. M., & Sarstedt, M. (2010). Management of multi-purpose stadiums: Importance and performance measurement of service interfaces. *International Journal of Services Technology and Management, 14*(2/3), 188–207.

Hwang, H., & Takane, Y. (2004). Generalized structured component analysis. *Psychometrika, 69*(1), 81–99.

Hwang, H., & Takane, Y. (2014). *Generalized structured component analysis: A component-based approach to structural equation modeling*. New York: Chapman & Hall.

Hwang, H., Ho, M.-H., & Lee, J. (2010). Generalized structured component analysis with latent interactions. *Psychometrika, 75*(2), 228–242.

Jöreskog, K. G. (1979). Basic ideas of factor and component analysis. In K. G. Jöreskog & D. Sörbom (Eds.), *Advances in factor analysis and structural equation models* (pp. 5–20). New York: University Press of America.

Jöreskog, K. G., & Wold, H. O. A. (1982). The ML and PLS techniques for modeling with latent variables: Historical and comparative aspects. In H. O. A. Wold & K. G. Jöreskog (Eds.), *Systems under indirect observation: Causality, structure, prediction, part I* (pp. 263–270). Amsterdam: North-Holland.

Kaufmann, L., & Gaeckler, J. (2015). A structured review of partial least squares in supply chain management research. *Journal of Purchasing and Supply Management, 21*, 259–272.

Kumar, B. R., & Waheed, K. A. (2015). Determinants of dividend policy: Evidence from GCC market. *Accounting and Finance Research, 4*(1), 17–29.

Lee, L., Petter, S., Fayard, D., & Robinson, S. (2011). On the use of partial least squares path modeling in accounting research. *International Journal of Accounting Information Systems, 12*, 305–328.

Lei, P.-W., & Wu, Q. (2007). Introduction to structural equation modeling: Issues and practical considerations. *Educational Measurement: Issues and Practice, 26*(3), 33–43.

Lohmöller, J. B. (1989). *Latent variable path modeling with partial least squares*. Heidelberg: Physica-Verlag.

Martensen, A., & Grønholdt, L. (2003). Improving library users' perceived quality, satisfaction and loyalty: An integrated measurement and management system. *The Journal of Academic Librarianship, 29*(3), 140–147.

Matthews, L., Sarstedt, M., Hair, J. F., & Ringle, C. M. (2016). Identifying and treating unobserved heterogeneity with FIMIX-PLS: Part II—A case study. *European Business Review, 28*(2), 208–224.

Monecke, A., & Leisch, F. (2012). semPLS: Structural equation modeling using partial least squares. *Journal of Statistical Software, 48*(3), 1–32.

Nitzl, C. (2016). The use of partial least squares structural equation Modelling (PLS-SEM) in management accounting research: Directions for future theory development. *Journal of Accounting Literature, 37*(December), 19–35.

Peng, D. X., & Lai, F. (2012). Using partial least squares in operations management research: A practical guideline and summary of past research. *Journal of Operations Management, 30*, 467–480.

Petter, S., Straub, D. W., & Rai, A. (2007). Specifying formative constructs in information systems research. *MIS Quarterly, 31*(4), 623–656.

Richter, N. F., Cepeda Carrión, G., Roldán, J. L., & Ringle, C. M. (2016a). European management research using partial least squares structural equation modeling (PLS-SEM): Editorial. *European Management Journal, 34*(6), 589–597.

Richter, N. F., Sinkovics, R. R., Ringle, C. M., & Schlägel, C. (2016b). A critical look at the use of SEM in International Business Research. *International Marketing Review, 33*(3), 376–404.

Rigdon, E. E. (2012). Rethinking partial least squares path modeling: In praise of simple methods. *Long Range Planning, 45*(5–6), 341–358.

Rigdon, E. E. (2014). Rethinking partial least squares path modeling: Breaking chains and forging ahead. *Long Range Planning, 47*(3), 161–167.

Rigdon, E. E. (2016). Choosing PLS path modeling as analytical method in European management research: A realist perspective. *European Management Journal, 34*(6), 598–605.

Rigdon, E. E., Sarstedt, M., & Ringle, C. M. (2017). On comparing results from CB-SEM and PLS-SEM. Five perspectives and five recommendations. *Marketing ZFP—Journal of Research and Management*. Forthcoming.

Ringle, C. M., & Sarstedt, M. (2016). Gain more insight from your PLS-SEM results: The importance-performance map analysis. *Industrial Management & Data Systems, 116*(9), 1865–1886.

Ringle, C. M., Sarstedt, M., & Straub, D. W. (2012). A critical look at the use of PLS-SEM in MIS quarterly. *MIS Quarterly, 36*, iii–xiv.

Ringle, C. M., Sarstedt, M., & Schlittgen, R. (2014). Genetic algorithm segmentation in partial least squares structural equation modeling. *OR Spectrum, 36*(1), 251–276.

Ringle, C. M., Wende, S., & Becker, J. M. (2015). *SmartPLS 3*. Bönningstedt: SmartPLS GmbH.

Rose, P. S. (1986). The bank branch: Which way to the future? *Canadian Banker, 93*(6), 42–50.

Sarstedt, M., Henseler, J., & Ringle, C. M. (2011). Multi-group analysis in partial least squares (PLS) path modeling: Alternative methods and empirical results. In M. Sarstedt, M. Schwaiger, & C. R. Taylor (Eds.), *Advances in international marketing* (Vol. 22, pp. 195–218). Bingley: Emerald.

Sarstedt, M., Ringle, C. M., Smith, D., Reams, R., & Hair, J. F. (2014). Partial least squares structural equation modeling (PLS-SEM): A useful tool for family business researchers. *Journal of Family Business Strategy, 5*(1), 105–115.

Sarstedt, M., Hair, J. F., Ringle, C. M., Thiele, K. O., & Gudergan, S. P. (2016). Estimation issues with PLS and CBSEM: Where the bias lies! *Journal of Business Research, 69*(10), 3998–4010.

Sarstedt, M., Ringle, C. M., & Hair, J. F. (2018). Partial least squares structural equation modeling. In C. Homburg, M. Klarmann, & A. Vomberg (Eds.), *Handbook of market research*. Heidelberg: Springer.

Schlittgen, R. (2017). Estimation of generalized structured component analysis models with alternating least squares. *Computational Statistics*. Forthcoming.

Schlittgen, R., Ringle, C. M., Sarstedt, M., & Becker, J.-M. (2016). Segmentation of PLS path models by iterative reweighted regressions. *Journal of Business Research, 69*, 4583–4592.

Schloderer, M. P., Sarstedt, M., & Ringle, C. M. (2014). The relevance of reputation in the nonprofit sector: The moderating effect of socio-demographic characteristics. *International Journal of Nonprofit and Voluntary Sector Marketing, 19*(2), 110–126.

Shmueli, G., Ray, S., Estrada, J. M. V., & Chatla, S. B. (2016). The elephant in the room: Evaluating the predictive performance of PLS models. *Journal of Business Research, 69*(10), 4552–4564.

Sohn, S. Y., Han, H. K., & Jeon, H. J. (2007). Development of an air force warehouse logistics index to continuously improve logistics capabilities. *European Journal of Operational Research, 183*, 148–161.

Tenenhaus, M., Amato, S., & Vinzi, V. E. (2004). A global goodness-of-fit index for PLS structural equation modeling. *Proceedings of the XLII SIS Scientific Meeting*, 739–742.

Tenenhaus, M., Vinzi, V. E., Chatelin, Y.-M., & Lauro, C. (2005). PLS path modeling. *Computational Statistics & Data Analysis, 48*, 159–205.

Teo, A. C., Tan, G. W. H., Ooi, K. B., & Lin, B. (2015). Why consumers adopt mobile payment? A partial least squares structural equation modelling (PLS-SEM) approach. *International Journal of Mobile Communications, 13*(5), 478–497.

Thiele, K.-O., Sarstedt, M., & Ringle, M. C. (2015). *Mirror, mirror on the wall. A comparative evaluation of new and established structural equation modeling methods*. 2nd International Symposium on Partial Least Squares Path Modeling, Seville, Spain.

Vinzi, E. V., Trinchera, L., & Amato, S. (2010). PLS path modeling: From foundations to recent developments and open issues for model assessment and improvement. In V. E. Vinzi, W. W. Chin, J. Henseler, & H. Wang (Eds.), *Handbook of partial least squares: Concepts, methods and applications: Vol. 2. Springer handbooks of computational statistics series* (pp. 47–82). Heidelberg: Springer.

Voorhees, C. M., Brady, M. K., Calantone, R., & Ramirez, E. (2016). Discriminant validity testing in marketing: An analysis, causes for concern, and proposed remedies. *Journal of the Academy of Marketing Science, 44*(1), 119–134.

Willaby, H. W., Costa, D. S. J., Burns, B. D., MacCann, C., & Roberts, R. D. (2015). Testing complex models with small sample sizes: A historical overview and empirical demonstration of what partial least squares (PLS) can offer differential psychology. *Personality and Individual Differences, 84*, 73–78.

Wold, H. O. A. (1982). Soft modeling: The basic design and some extensions. In K. G. Jöreskog & H. O. A. Wold (Eds.), *Systems under indirect observations: Part II* (pp. 1–54). Amsterdam: North-Holland.

Wold, H. O. A. (2006). *Partial least squares encyclopedia of statistical sciences*. New York: Wiley.

Wu, W. W., Lan, L. W., & Lee, Y. T. (2012). Exploring the critical pillars and causal relations within the NRI: An innovative approach. *European Journal of Operational Research, 218*, 230–238.

Chapter 2
Bank Soundness: A PLS-SEM Approach

Charmele Ayadurai and Rasol Eskandari

Abstract During the Global Financial Crisis (GFC) of 2007–2009, even banks in industrial economies with long established markets suffered significantly. This highlights, weaknesses in the banking system and the importance of a sound banking sector. This paper applies Partial Least Squares Structural Equation Modeling (PLS-SEM) to explain the drivers of bank soundness in the G7 countries during the period 2003–2013. PLS-SEM models are able to handle latent variables and complex models, and thus, PLS-SEM is suitable for this study. In creating a parsimonious model, the study assembles 17 manifest variables of six constructs as the direct cause and eight constructs as the indirect cause of bank soundness. The structural equation model comprises of six latent exogenous constructs [Capital (C), Asset (A), Management (M), Earnings (E), Liquidity (L) and Sensitivity (S)] which explains the observed consequences of bank soundness in these countries. Results indicate that CAMELS constructs are able to explain 32.5% of the variation in banks' soundness. The model's predictive relevance (Q^2) in regards to endogenous construct stands at a medium category of 0.315. The results imply that banks placed high importance on off-balance sheet and capital activities, and thus, taking on higher risk. Surprisingly, banks were also operating at low levels of capital and liquidity, resembling banks that failed during the Great Depression of the 1930s. The weakness in capital and liquidity measures shows the need for policy makers to have a better understanding of sound banking, before quantifying measures and creating policies that makes banks' less prone to crises episodes and create convergence with soundness.

Keywords Bank soundness · G7 · CAMELS · Financial crises · Partial least squares

C. Ayadurai (✉) • R. Eskandari
School of Business and Law, University of Salford, Manchester, UK
e-mail: c.ayadurai@edu.salford.ac.uk; R.Eskandari@salford.ac.uk

N.K. Avkiran, C.M. Ringle (eds.), *Partial Least Squares Structural Equation Modeling*, International Series in Operations Research & Management Science 267, https://doi.org/10.1007/978-3-319-71691-6_2

2.1 Introduction

The recent GFC resulted in large bank failures in the G7 countries (Canada, France, Germany, Italy, Japan, UK and US) with the exception of Canada, which led to deep recessions. This highlights the chronic weakness in the banking sector and the importance of a robust banking system. A sound banking system has to be efficient in the key role it plays to influence the economy positively.

This article assesses the soundness of banks in G7 countries, to understand the workings of these banks as providers of payment services and hubs for economic and financial activities during the crisis period. The motivation to study the G7 banks lies on several reasons. G7 countries are among the top ten biggest economies, five of the countries are the top ten financial hubs in the world and G7 countries play a key role in global monetary affairs and trade. As a result, these countries hold a significant position in influencing the world economy at large.

In assessing the soundness of banks, seven clusters of financial indicators and credit rating models were considered, namely, Financial Soundness Indicator (FSI), Basel Core Principles (BCP) , CAMELS, Moody, Fitch, Standard and Poor (S&P) and Bank Financial Strength Rating (BFSR). The study puts together 60 bank-level variables from these indicators and applies stock returns as the proxy for bank soundness. Partial least squares structural equation modeling (PLS-SEM; e.g., Haenlein and Kaplan 2004; Reinartz et al. 2009; Hair et al. 2017) assist to assess the banks.

The study offers two main contributions. First, the assessment of G7 banks as key players in global trade and monetary affairs during the GFC. Second, contribution is to apply seven sets of bank soundness and credit rating indicators in the assessment.

2.2 Literature Review

Bank soundness is a concept that signifies the ability of a bank to survive an adversity in the economy (Lindgren et al. 1996). Financial ratios play a key role in assessing bank soundness, as early signs of impairment could be easily detected by the changes in the internal condition of the banks (Sinkey 1979; Hanc 1998). On top of that, Avkiran et al. (forthcoming) suggest that microlevel linkages play a major role in the contagion of systemic risk. Several measures have been suggested for bank soundness such as earnings (Gasbarro et al. 2002); capital (Schaeck and Cihak 2012); internal governance (Lindgren et al. 1996; Barr and Siems 1997) and credit ratings (Podviezko and Ginevičius 2010; Demirgüç-Kunt and Detragiache 2011). A sound bank is a bank that is solvent and remains solvent. The future solvency of a bank depends on its efficiency, and thus, its profitability. Therefore, solvency is a measure of the positive net worth of a bank.

As banks go through different phases, it is impossible to classify banks, precisely as "sound" or "unsound" at a given point of time. This is because banks could be performing well at the moment but show signs of probable problems in the future. As theory is unable to provide a clear answer on what constitutes a sound bank, the current study looks at bank specific sets of indicators for solutions.

In literature as reported by Bernanke (2007), the Fed examines the safety and soundness of banks in US through CAMELS ratings. CAMELS' ratings proved to be effective in reflecting bank soundness (Meyer and Pifer 1970; Korobow and Stuhr 1975; Korobow et al. 1976, 1977; Pettway and Sinkey 1980; Bovenzi et al. 1987). Basel Core Principles (BCP) on the other hand, represents the global standard for best practices in supervision and regulation. Demirgüç-Kunt and Detragiache (2011) noted that better compliance with BCP does not necessarily lead to improved bank soundness. Models such as Financial Soundness Indicators (FSI) "core" series, monitors the soundness of the banks as Authorised Deposit-taking Institutions (ADI) (Costa Navajas and Thegeya 2013).

Demirgüç-Kunt et al. (2008) state that rating measures bank soundness accurately as credit rating system takes into account both quantitative and qualitative information on banks as well as the environment. Moody denotes that BFSR (Bank Financial Strength Rating) measures the intrinsic soundness and safety of a bank. Moody, Fitch and Standard & Poor (S&P) measures the ability of banks to meet their depositors and creditors obligations as they fall due.

The study assembles seven sets of bank and credit rating indicators [Basel Core Principles (BCP) , Financial Soundness Indicators (FSI), CAMELS, Moody, Fitch, S&P and Bank Financial Strength Rating BFSR] to provide a meaningful insight on the indicators that contribute to bank soundness. Although several studies have assessed bank soundness in the past, only a few covered the two key areas of developed economies and soundness indicators (Gaganis et al. 2006; Demirgüç-Kunt et al. 2008; Demirgüç-Kunt and Detragiache 2011).

The core hypotheses tested through PLS-SEM are as follows:

H1: Capital makes a significant contribution to bank soundness
The quality and level of a bank's capital determines the survival of a bank. When a bank's capital buffer increases, it absorbs imbalances at both micro and macro levels (Roman and Şargu 2013) and acts as the last resort to uninsured depositors, creditors and the Federal Deposit Insurance Corporation (FDIC). Therefore, banks with higher levels of buffer are sound and conversely insufficient capital during adversity could bring banks down. However, higher buffers result in lesser proportion of capital allocated for investment purposes. This results in banks holding lower quality assets. As equity holders, receive lesser returns on their investments earnings decreases, and thus, has an adverse effect on bank soundness. In conclusion, capital, assets and earnings establishes a link with bank soundness.

H2: Asset makes a significant contribution to bank soundness
Banks that extend loans to credit worthy customers with sound collateral levels show low non-performing loans (NPLs) and less exposure to excessive risk levels. Therefore, higher quality assets results in higher profit margins, and thus, increases earnings. However, if banks have accumulated high NPLs, bad debts, and do not have quality collateral to back its loans, then there is a lesser chance of survival, and thus, establishing a link between assets, earnings and bank soundness.

H3: Management makes a significant contribution to bank soundness
The efficiency of the management structure lies in the ability of bank officers and managers to make decisions that contributes to bank soundness. Management's efficiency is usually reflected in the earnings generated (returns on assets and equity) and how effectively profit margins are maintained and increased. This establishes the link between management, earnings and bank soundness.

H4: Earnings makes a significant contribution to bank soundness
Earnings are first line of defense against adversity and loan defaults. If banks have accumulated large NPLs, profit margin and the quality of assets reduces. Therefore, steady streams of earnings from solid operating base is vital for the survival of the banks, and thus, establishing a link between earnings and bank soundness.

H5: Liquidity makes a significant contribution to bank soundness
Banks' funding sources such as corporate bonds, deposits and liquid assets such as cash in hand determine the ability of the bank to meet unforeseen deposit outflows. Banks that are unable to meet its daily liquidity needs could result in bank runs, and thus, leading to an insolvent bank. Liquidity inversely relates to earnings. This establishes a relationship between liquidity, earnings and bank soundness.

H6: Sensitivity makes a significant contribution to bank soundness
Market forces play a key role in bank stability. Banks exposes itself to various market risks (interest rate risk, foreign exchange risk, price risk). Banks are also sensitive to growth. Bank size (large or small measured in total assets) plays a key role in bank soundness (Bell 1997; Hooks 1995; Ohlson 1980; Gunsel 2005; Nurazi and Evans 2005). Lager banks have better access to finance, risk diversification and dealing with liquidity problems. As a result, big banks are more stable and financially sound, and thus, have longer survival time and are less likely to fail in comparison to smaller banks (Gunsel 2005). This finding was evident during the GFC (Košak et al. 2015). Therefore, the study assesses the effect of sensitivity ratios such as sensitivity to market risk and bank size ratios, on asset, earnings and capital of a bank.

2.3 Data Collection

The sample consists of 1135 listed banks in G7 countries. Listed banks provide homogeneity in the comparison of banks with different economies. Banks chosen were under the Global Industry Classification Standard of banks (code 401010) in Osiris database. The study collects data for the period 2003–2013 sourced from Osiris and Bankscope databases. The data was converted and averaged in US dollars.

The study collected 60 independent variables (see Tables 2.1 and 2.2) and categorized them under CAMELS. In reference to Fama and French (1992), stock returns explain microeconomic variables such as CAMELS and is forward looking. Thus, is able to measure banks' expected soundness.

Summary statistics highlights substantial skewness and kurtosis outside the range of ±2.58 across the variables, and thus, failing to meet parametric assumptions (details are available from the author). According to Cheng (2008), King and Wen (2011), and Rasli et al. (2013) archive-based financial accounting empirical studies often report non-normal datasets.

2.4 Methodology

The study selects a second-generation multivariate technique of PLS-SEM for several reasons. The PLS-SEM model takes care of measurement errors in variables (Chin 1998). Accounting and finance discipline applies PLS-SEM in their research (Lee et al. 2011; Nitzl 2016; Avkiran et al. forthcoming). As PLS-SEM is a non-parametric estimator, it is the most appropriate model for this research. The goal of the research is to identify key drivers amongst the CAMELS constructs that determines bank soundness. PLS-SEM helps in this respect to detect if a relationship exists between variables while confirming the theory that underlines the variable (Chin 1998), which is a vital contributor to this study. On top of that, Richter et al. (2015) state that PLS-SEM conducts a proficient analysis.

As all the 60 manifest variables are formative, PLS-SEM works well with complex models that incorporates formative measures in answering the research question. Bank specific variables of CAMELS contain observable variables such as the ratios and latent factors that are not directly observable. One example is the variables under the earnings cluster "return on average assets" and "return on average equity". These variables are highly correlated among themselves but have small correlations with Capital, Asset, Management, Liquidity and Sensitivity variables. This suggests the presence of latent variable "profitability" or "earnings" in Capital, Asset, Management and Liquidity variables, which are responsible for the observed correlations.

PLS-SEM is useful in explaining the maximum variance of a latent construct (Avkiran et al. forthcoming). As PLS-SEM integrates both econometric and psycho-

Table 2.1 List of independent variables

Category	Code	Variable
Asset	aa	Allowance for loan loss to gross loan
	ab	Common equity to net loan
	ac	Equity to net loan
	ad	Gross non-performing loan to advances
	ae	Non-performing loan to gross loan
	af	Provision for loan loss to net advances
	ag	Loan loss provision to average asset
	ah	Loan loss provision to net interest income
	ai	Provision for loan loss to total loan
	aj	Non-performing loan to net advances
	ak	Non-performing loan to total equity
	al	Total loan to total asset
Capital	ca	Capital adequacy ratio 1
	cb	Capital adequacy ratio 2
	cc	Common equity to total asset
	cd	Debt to equity
	ce	Retained earnings to total equity
	cf	Net income and total equity to deposit and short term funding
	cg	Net income and total equity to total asset
	ch	Regulatory tier 1 capital to risk weighted asset
	ci	Equity to asset
Management	ma	Business per employee to total shareholder return
		Loan growth rate
	mb	Management expense to average asset
	mc	Profit per employee to total shareholder return
	md	Total loan to total deposit
Earnings	ea	Cost to income
	eb	Dividend payment to net income
	ec	Earnings per share to average equity
	ed	Interest income to interest expense
	ee	Interest income to total income
	ef	Non-interest expense to average asset
	eg	Non-interest expense to gross income
	eh	Non-interest expense to total expense
	ei	Non-interest expense to total customer deposit
	ej	Net interest margin to gross income
	ek	Net interest margin
	el	Net interest revenue to average asset
	em	Non-interest income to total income

(continued)

Table 2.1 (continued)

Category	Code	Variable
	en	Net income to average asset
	eo	Net interest income to asset growth ratc
	ep	Non-interest income to non-interest expense
	eq	Operating income to total asset
	er	Pre-tax income to average asset
	es	Pre-tax income to revenue
	et	Net profit to average asset
	eu	Net profit on average equity
	ev	Tax to earning before tax
	ew	Interest expense to total expenses
Liquidity	la	Customer deposit to total asset
	lb	Liquid asset to customer and short term funding
	lc	Liquid asset to deposit and non-deposit fund
	ld	Liquid asset to short term liabilities
	le	Liquid asset to total asset
	lf	Liquid asset to total deposit
	lg	Net loan to total asset
	lh	Non-performing loan to total asset
	li	Total loan to customer deposit
Sensitivity	sa	Log of total asset
	se	Market price per ordinary equity to earning per share
	sc	Log of total asset + market price per ordinary equity to earning per share

This table reports 60 independent variables to analyse bank soundness. All the variables are in the form of financial ratios. Results are in percentage form. These variables are categorised under C (capital), A (asset), M (management), E (earning), L (liquidity) and S (sensitivity). The dataset is from Bankscope and Osiris database

metric analysis in its estimation (Fornell and Larcker 1981; Wold 1985; Chin 1998), it is the appropriate measure for both observable and unobservable variables in empirical studies. In this respect, PLS-SEM works to understand the relationship among the variables by understanding the constructs that underlie them and how the latent factors drive the variation in the data. On PLS-SEM historical background, refer to Chin (1998), Lohmöller (1989), Dijkstra (2010, 2014), and Rigdon (2012, 2014).

The initial model design had six exogenous constructs (CAMELS) with 60 manifest variables. In creating a parsimonious model, the exogenous constructs were remodeled to six direct constructs, eight indirect constructs and manifest variables were reduced to 17 [2 capital (C), 5 assets (A), 2 management (M), 3 earnings (E), 3 liquidity (L), 1 sensitivity (S)]. The choice of measurement mode follows the work of Bastan et al. (2016). In reference to Fig. 2.1, path models visually show the relationships between the six hypotheses and manifest variables (Hair et al. 2011, 2017). The inner model (structural model) displays the relationships between

Table 2.2 List of variables with references

No	Code	Variables	References
	Dependent variable	*Stock returns*	Fama and French (1992)
	Independent variables		
Asset			
1	aa	Allowance for loan loss to gross loan	Dang (2011)
2	ab	Common equity to net loan	Poon et al. (1999)
3	ac	Equity to net loan	Poon et al. (1999)
4	ad	Gross non-performing loan to advances	Kumar et al. (2012)
5	ae	Nonperforming loans to gross loans	International Monetary Fund Staff (2008) and Podviezko and Ginevičius (2010)
6	af	Provision for loan loss to net advances	Kumar et al. (2012)
7	ag	Loan loss provision to average asset	Poon et al. (1999)
8	ah	Loan loss provision to net interest income	Loannidis et al. (2010) and Gaganis et al. (2006)
9	ai	Provision for loan loss to total loans	Kumar et al. (2012)
10	aj	Non-performing loan to net advances	Kumar et al. (2012)
11	ak	Non-performing loan to total equity	Dang (2011)
12	al	Total loan to total asset	Kumar et al. (2012)
Capital			
13	ca	Capital adequacy ratio(1)	Dang (2011)
14	cb	Capital adequacy ratio(2)	Kumar et al. (2012)
15	cc	Common equity to total asset	Poon et al. (1999)
16	cd	Debt to equity	Toor (2006)
17	ce	Retained earnings to total equity	Sarker (2006)
18	cf	Net income and total equity to deposit and short term funding	Canbas et al. (2005)
19	cg	Net income and total equity to total asset	Canbas et al. (2005)
20	ch	Regulatory Tier 1 capital to risk weighted assets	International Monetary Fund Staff (2008) and Podviezko and Ginevičius (2010)
21	ci	Total equity to total asset	Loannidis et al. (2010), Gaganis et al. (2006) and Podviezko and Ginevičius (2010)

(continued)

Table 2.2 (continued)

No	Code	Variables	References
Management			
22	ma	Business per employee to total shareholder return	Kumar et al. (2012)
23		Loan growth rate	Dang (2011) and Podviezko and Ginevičius (2010)
24	mb	Management expense to average asset	Kumar et al. (2012)
25	mc	Profit per employee to total shareholder return	Kumar et al. (2012)
26	md	Total loan to total deposit	Kumar et al. (2012)
Earnings			
27	ea	Cost to income	Dang (2011) and Gaganis et al. (2006)
28	eb	Dividend payment to net income	Poon et al. (1999)
29	ec	Earning per share to average equity	Toor (2006)
30	ed	Interest income to interest expense	Canbas et al. (2005)
31	ee	Interest income to total income	Toor (2006)
32	ef	Non-interest expenditure to average asset	Loannidis et al. (2010)
33	eg	Non-interest expense to gross income	International Monetary Fund Staff (2008) and Podviezko and Ginevičius (2010)
34	eh	Non-interest expenses to total expense	International Monetary Fund Staff (2008)
35	ei	Non-interest expense/total deposit	Poon et al. (1999)
36	ej	Interest margin to gross income	International Monetary Fund Staff (2008) and Podviezko and Ginevičius (2010)
37	ek	Net interest margin	Poon et al. (1999)
38	el	Net interest revenue to average asset	Dang (2011)
39	em	Non-interest income to total income	Toor (2006)
40	en	Net income to average asset	Poon et al. (1999), Gaganis et al. (2006) and Podviezko and Ginevičius (2010)
41	eo	Net interest income to asset growth rate	Dang (2011)
42	ep	Non-interest income to non-interest expense	International Monetary Fund Staff (2008)
43	eq	Operating profit to total asset	Gasbarro et al. (2002)
44	er	Pre-tax income to average asset	Poon et al. (1999)

(continued)

Table 2.2 (continued)

No	Code	Variables	References
45	es	Pre-tax income to revenue	Poon et al. (1999)
46	et	Net profit to average asset	Kumar et al. (2012)
47	eu	Net profit on average equity	Demirgüç-Kunt et al. (2008)
48	ev	Tax to earning before tax	Poon et al. (1999) and Podviezko and Ginevičius (2010)
49	ew	Interest expense to total expenses	Canbas et al. (2005)
Liquidity			
50	la	Customer deposits to total assets	Dang (2011)
51	lb	Liquid asset to customer and short term funding	Loannidis et al. (2010) and Gaganis et al. (2006)
52	lc	Liquid asset to deposit and non-deposit fund	Canbas et al. (2005)
53	ld	Liquid asset to short term liabilities	International Monetary Fund Staff (2008)
54	le	Liquid asset to total asset	Poon et al. (1999)
55	lf	Liquid assets to total deposit	Kumar et al. (2012)
56	lg	Net loan to total asset	Demirgüç-Kunt et al. (2008)
57	lh	Non-performing loan to total asset	Gasbarro et al. (2002)
58	li	Total loan to customer deposit	Dang (2011)
Sensitivity			
59	sa	Log of total asset	Košak et al. (2015) and Gaganis et al. (2006)
60	se	Market price per ordinary equity to earning per share	Nurazi and Evans (2005)

This table reports the source of article for the 60 independent variables and 1 dependent variable. These variables were collected through reviewing 50 journal articles from 1968 to 2017. The variables source is from seven clusters of financial indicators and credit rating models addressed by the following authors. Financial Soundness Indicator (FSI)—International Monetary Fund Staff (2008); Basel Core Principles (BCP)—Demirgüç-Kunt et al. (2008), CAMELS—Dang (2011); Kumar et al. (2012); Sarker (2006); Canbas et al. (2005) and Gasbarro et al. (2002); Moody—Poon et al. (1999); Fitch—Loannidis et al. (2010); Gaganis et al. (2006); Standard and Poor (S&P)—Podviezko and Ginevičius (2010) and Bank Financial Strength Rating (BFSR) —Poon et al. (1999)

constructs. While the outer model (measurement model) displays the relationship between the constructs and the manifest variables. Henseler et al. (2009) notes that the partial least squares path model comprises of inner and outer models, which sets out the linear equation. The inner model highlights the relationships amongst the latent constructs.

The inner model construction is as follows:

$$\xi = B\xi + \zeta \tag{2.1}$$

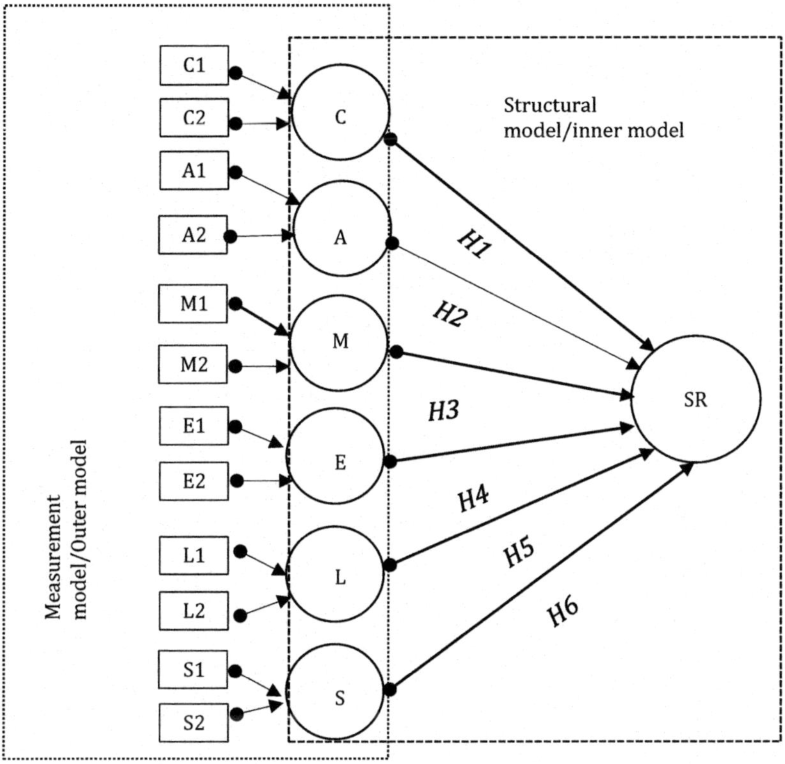

Fig. 2.1 Structural model/inner model and the measurement model/outer model

where ξ stands for the vector of latent variables, B is the matrix of coefficients and ζ is the inner model residuals. The predictor specification reduces the inner model in Eq. (2.1) to:

$$\left(\xi|\xi\right) = B\xi \tag{2.2}$$

The outer model predicts the relationship amongst the latent constructs and indicators. The outer model comprises of reflective and formative measurement models.

The reflective mode shows causal relationship from latent variable to manifest variables generated as a linear function of latent variables and residual ε:

$$X_{\chi} = \Lambda_{\chi}\xi + \varepsilon_{\chi} \tag{2.3}$$

where Λ stands for the loading coefficients. The predictor specification reduces the outer model in Eq. (2.3) to:

$$\left(X_\chi | \xi\right) = \Lambda_\chi \xi \tag{2.4}$$

The formative shows causal relationship from manifest variables to latent variables. The linear relationship is as follows:

$$\xi = \Pi_\chi X_\chi + \varepsilon_\chi \tag{2.5}$$

The predictor specification reduces Eq. (2.5) to:

$$\left(\xi | X_\chi\right) = \Pi_\chi X_\chi \tag{2.6}$$

For more explanations of the algorithm, please refer to Wold (1982), Tenenhaus and Vinzi (2005) and Rigdon (2014). PLS-SEM algorithm comprises of two stages. The first stage estimates the latent constructs' scores in four steps. The study first considers the outer approximations of latent construct scores. Then proxies for structural model relationships are established. The study then considers the inner approximation of latent constructs scores while estimating the proxies for coefficients in the measurement models. The second stage calculates the final estimates of the outer weights and loadings and path coefficients (Lohmöller 1989).

2.5 Findings

Theoretical conceptualization supports the framework that CAMELS constructs are appropriate measures of stock returns, and thus, the proxy for bank soundness. The 17 manifest variables form the six direct CAMELS exogenous constructs. These constructs model as formative measures for the endogenous construct of stock returns (Fig. 2.2). The study follows Hair et al. (2017) in assessing the formative measures. Formative measures are free of errors (Diamantopoulos 2006; Edwards and Bagozzi 2000). The study presents a comprehensive set of formative indicators (Table 2.1) to show that the formative indicators encapsulate all the facets of the construct. PLS-SEM applies bootstrapping (Davison and Hinkley 1997; Efron and Tibshirani 1993) which is a repetitive random sampling to obtain standard errors for hypothesis testing. This process allows the significance of the coefficients to be tested (Henseler et al. 2009). The bootstrapping procedure generates outer weights, outer loadings and path coefficients results. The bootstrapping procedure ran with a sample of 10,000 with "no sign change" option for the most conservative results (Table 2.3).

In assessing the significance and the relevance of the formative indicators, the study examines the outer weights. Table 2.4 assembles the relative contribution of each manifest variables (in weights), and thus, its significance in forming the constructs. The results show that some manifest variables have low or insignificant outer weights. Although the outer weight is insignificant, the outer loading is above 0.5. This retains the construct as the manifest variables have an absolute contribution to the constructs.

Variance Inflation Factor (VIF) points out the collinearity issues in formative measurements models (see Table 2.4). The results show that all VIFs are below the threshold level of five. Therefore, the presence of collinearity issues in manifest variables is not of concern. The study managed to create a parsimonious model with a R^2 value of 32.5%. This indicates that the direct CAMELS constructs explain 32.5% of the variance in stock returns. Hair et al. (2011) and Henseler et al. (2009) evaluates this result as moderate. While the indirect constructs, asset and capital explain the least variance in stock returns (0.010; 0.000). Earnings with a strong R^2 value of 0.832 explains the most variance in stock returns in this study. The model's predictive relevance (Q^2) with regard to endogenous construct stands at a moderate category of 0.315 for stock returns (see Hair et al. 2017). The indirect constructs of asset and capital have a low predictive relevance of 0.004 and 0.000, while earnings have a moderate predictive relevance of 0.233.

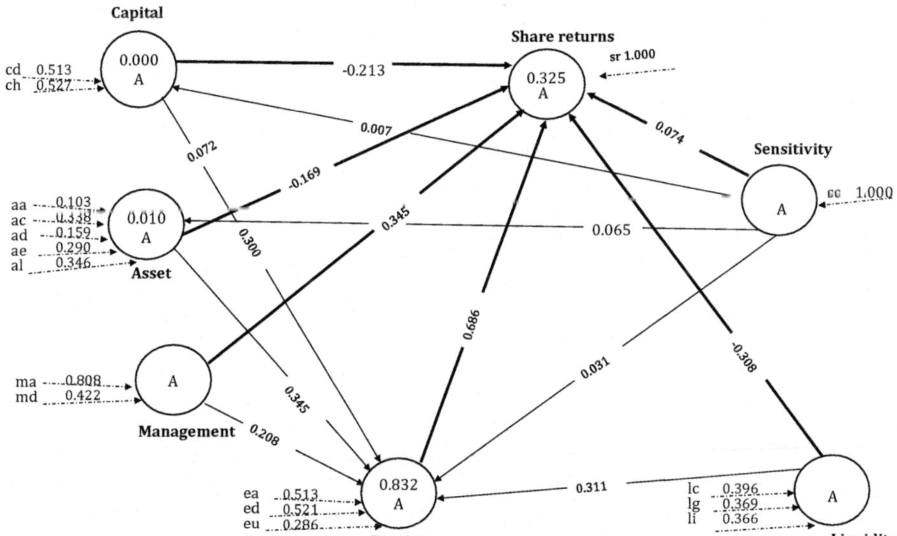

Fig. 2.2 PLS-SEM results

⟶	**Effect of direct constructs (CAMELS) on stock returns** Capital on stock returns Asset on stock returns Management on stock returns Earnings on stock returns Liquidity on stock returns Sensitivity on stock returns
⟶	**Effect of indirect constructs** Capital on earnings Capital on asset Asset on earnings Management on earnings Liquidity on earnings Sensitivity on capital Sensitivity on earnings Sensitivity on asset
-·-·-·-·→	**Effect of formative manifest variableson direct constructs** aa on Asset ac on Asset ad on Asset ae on Asset al on Asset cd on Capital ch on Capital ea on Earnings ed on Earnings eu on Earnings lc on Liquidity lg on Liquidity li on Liquidity ma on Management md on Management sc on Sensitivity sr on Stock returns

Fig. 2.2 (continued)

2.6 Analysis

Capital construct failed to establish a significant relationship at 95% confidence level with stock returns (−0.038, 0.576) and earnings (0.010, 0.254). Capital has no predictive relevance as f square stands at a small level of 0.015 for stock returns, 0.005 for asset and 0.136 for earnings. Capital has also weak path coefficients of 0.072 for asset, 0.300 for earnings and −0.213 for stock returns. Thus, we reject hypothesis 1. The negative sign indicates insufficient capital levels, despite adhering to Basel Core Principles. The findings validate both Demirgüç-Kunt et al. (2008)

Table 2.3 Results for bank soundness constructs

Bank soundness constructs	Path coefficients	95% BCa confidence interval	f square	Significant?
Test criterion			Small 0.02	
			Medium 0.15	
			Large 0.35	
Capital—stock returns	−0.213	[−0.034, 0.600]	0.015	No
Capital—earning	0.300	[0.009, 0.251]	0.136	No
Capital—asset	0.072		0.005	
Asset—stock returns	−0.169	[0.123, 0.355]	0.012	Yes
Asset—earnings	0.345		0.259	
Management—stock returns	0.345	[0.096, 0.206]	0.074	Yes
Management—earnings	0.208		0.122	
Earnings—stock returns	0.686		0.117	
Liquidity—stock returns	−0.308	[0.140, 0.367]	0.021	Yes
Liquidity—earnings	0.311		0.094	
Sensitivity—stock returns	0.074	[0.007, 0.056]	0.008	No
Sensitivity—capital	0.007		0.000	
Sensitivity—earnings	0.031	[0.008, 0.060]	0.006	No
Sensitivity—asset	0.065	[−0.002, 0.014]	0.004	No

This table is a summary of the direct and indirect formative measured constructs' results. The table displays path coefficients, 95% confidence interval, 95% Bias Corrected confidence interval, f square and significance results

Table 2.4 Results for bank soundness manifest variables

Bank soundness manifest variables	Outer weights	VIF	95% Bca confidence interval	Significant?
Test criterion		<5		
Aa-Asset	0.103	1.762	[0.041,0.152]	Yes
Ac-Asset	0.338	1.901	[0.308,0.365]	Yes
Ad-Asset	0.159	3.501	[0.102,0.185]	Yes
Ae-Asset	0.29	3.82	[0.257,0.329]	Yes
Al-Asset	0.346	2.722	[0.320,0.385]	Yes
Cd-Capital	0.513	3.52	[0.414,0.721]	Yes
Ch-Capital	0.527	3.52	[0.486,0.909]	Yes
Ea-Earnings	0.513	1.349	[0.287,0.637]	Yes
Ed-Earnings	0.521	1.895	[0.482,0.561]	Yes
Eu-Earnings	0.286	1.555	[0.045,0.444]	Yes
Lc-Liquidity	0.396	2.969	[0.370,0.440]	Yes
Lg-Liquidity	0.369	1.925	[0.333,0.418]	Yes
Li-Liquidity	0.366	2.311	[0.334,0.389]	Yes
Ma-Management	0.396	1.066	[0.754,0.855]	Yes
Md-Management	0.369	1.066	[0.361,0.485]	Yes

This table summarises the formative measured manifest variables' results. The table displays outer weights and VIF results and Berger and Bouwman (2009) points that better capitalization only leads to sound banks in the early 1990s but not during the recent GFC. Thus, mere adherence to Basel Core Principles does not guarantee bank soundness. In the effects of asset on stock returns, the findings highlight that asset (0.132, 0.366) has established a strong significance at 95% confidence level. Asset has an average predictive relevance (f square 0.259) for earnings but a weak predictive relevance of 0.012 for stock returns. In terms of path coefficients, a weak value of −0.169 is evident for stock returns and 0.345 for earnings. Thus, supporting hypothesis 2. The path coefficients highlight a negative relationship. This suggests that these countries had low levels of assets.

Management construct shows significance with stock returns (0.090, 0.197) at 95% confidence level. Thus, establishing a relationship between the constructs and supporting hypothesis 3. However, management has no predictive relevance with an f square of 0.074 for stock returns, 0.122 for earnings and weak but positive path coefficients of 0.345 (stock returns) and 0.208 (earnings). This suggests that management is less important in determining bank soundness. On the contrary, Lindgren et al. (1996) and Barr and Siems (1997) view internal governance (management) as the most important construct for a sound bank. A weak predictive relevance is evident with an f square that stands at 0.117. The path coefficients display a value of 0.686 suggesting that earnings have a moderate and a positive relationship with stock returns, and thus, supporting hypothesis 4.

Liquidity construct highlights a weak path coefficients of −0.308 for stock returns and 0.311 for earnings and no predictive relevance (f square = 0.021 stock returns), (f square = 0.094 earnings) but a strong significance at 95% confidence level with stock returns (0.099, 0.297). Therefore, supporting hypothesis 5. However, the path coefficients show an adverse relationship with stock return suggesting that banks had insufficient liquidity buffer. Low liquidity levels result in dangerous bank runs. Ratnovski and Huang (2009) on the other hand, noted that banks in UK and US were relatively liquid during the GFC.

Sensitivity have weak path coefficients with capital (0.007), earning (0.031), asset (0.065) and stock returns (0.074). In terms of significance, none of the constructs [asset (0.000, 0.021), stock returns (0.009, 0.059) and earnings (0.011, 0.072)] was significant at 95% confidence level. Thus, not supporting hypothesis 6. The predictive relevance were weak [f square = 0.004 (asset), 0.008 (stock returns), 0.000 (capital) and 0.006 (earnings)]. The results suggest that sensitivity has less effect on bank soundness. This finding contradicts Demirgüç-Kunt et al. (2008) findings that big banks diversify, and thus, are more stable and sound. However, it is in line with Gaganis et al. (2006) findings that bank size does not determine bank soundness in developed countries. Ho and Saunders (1980) suggested that large banks have access to discount windows and have partially insured depositors. Thus, are more likely to have catastrophic consequences than smaller banks.

The summarized results, (see Table 2.5 and Fig. 2.3) suggests that earnings plays a significant role while sensitivity is insignificant in determining bank soundness. Banks were also least focused on core business areas of taking deposits and giving out loans. Off-balance sheet activities and capital market investment have higher priority, and thus, banks were taking on excessive risks. This suggests that managements' gamble to take on high risk paid off with a moderate return on earnings. However, failing to manage and cope with risk, led to disastrous impact on capital, asset and liquidity ratios.

Surprisingly, banks were also operating with insufficient capital and liquidity ratios which resembles the causes of bank failures (illiquidity, bad assets, overbanking and mismanagement) during the Great Depression of 1930s (Tussing 1967). Banks engage in various form of financing such as long-term assets, fund short-term debts and carry out excessive amounts of maturity transformations. Aggressive withdrawal of funds during adversity could cause bank runs. Although Basel III established Net Stable Funding Ratio (NFSR) and Liquidity Coverage Ratio (LCR) with the intention to improve liquidity levels, the results from this study indicate otherwise. Despite the fact that capital is a strong determinant of bank status, capital showed no significance in bank soundness. As banks grow in size, capital needs to increase proportionately to cushion against market shocks. Basel III forms the understanding that higher bank capital results in financial stability. Conversely, Ratnovski and Huang (2009) and Währungsfonds (2009) found that banks with high capital levels in advanced economies exhausted capital to adversity during the GFC. Therefore, there is no conclusive answer whether increased capital levels helps during crisis periods.

2.7 Conclusions

The study demonstrates how PLS-SEM parsimoniously models bank soundness decisions. The results suggest that the focus of banks in G7 countries were on off-balance sheet transactions and capital market investments, which led to higher risk levels. These banks also operated with insufficient capital and liquidity to cushion adversity, resulting in bank failures. This resembles Bakir's (2013) finding that banks take excessive risk because of inadequate financial regulations and supervision.

In reference to Calomiris and Herring (2013), Basel III does not provide solutions to two core problems, risk measurement and replacement of lost capital in a timely manner. Since book capital ratios and supervision proved ineffective during the crisis, Calomiris and Herring (2013) suggest great emphasis on a well-designed convertible contingent capital (CoCos) to improve incentives for risk management and maintaining adequate capital for large banks.

Kodres and Narain (2010) suggested several guidelines for banks in the wake of GFC. Shock absorbers enhancers such as increasing capital and liquidity buffers while constraining leverage. While Moosa (2010) suggests adherence to risk management framework is vital, where liquidity and leverage are positioned as superior constructs to capital. As size is often seen as the most significant measure of a financial institution, a diversion to refocus on leverage, interconnectedness and complexity aspects should be implemented. Limits on size and scope of bank growth would also mitigate the "too big to fall" problem.

Table 2.5 Summary table

	Path coefficients	95% confidence interval	f square
Capital—stock returns	▼●	○	○
Capital—earning	●	○	○
Capital—asset	○	✚	○
Asset—stock returns	▼○	✚	○
Asset—earnings	●	★	✚
Management—stock returns	●	★	○
Management—earnings	●	★	○
Earnings—stock returns	✚	★	○
Liquidity—stock returns	▼●	★	○
Liquidity—Earnings	●	★	○
Sensitivity—stock returns	○	○	○
Sensitivity—capital	○	○	○
Sensitivity—earnings	○	○	○
Sensitivity—asset	○	○	○

Note: ○ = very weak, ● = weak, ▼ = negative, ✚ = moderate, ★ = exceptional
The table displays the overall summary results of direct and indirect constructs in symbols. The results presented here are path coefficients, confidence levels and f square

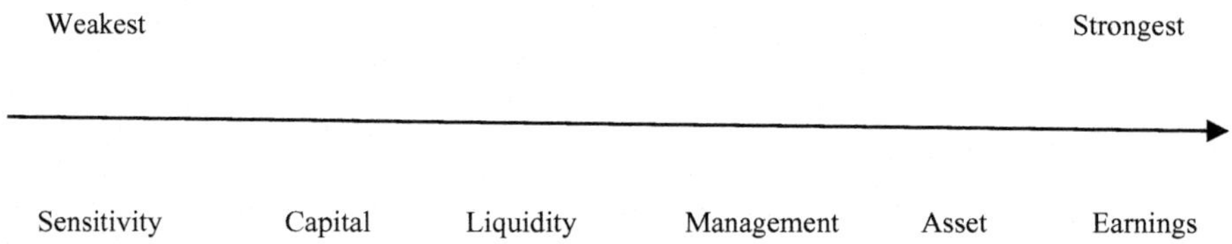

Fig. 2.3 Results of the weakest to the strongest construct

Procyclicality on the other hand, could be dampened by changes in several areas namely rules on accounting practices, risk management systems, capital and liquidity regimes, provisioning for losses and compensation schemes. In managing bank runs and panics, banks should have an underpinning confidence in the financial system's continuity during bankruptcy or liquidity. Moosa (2010) stresses the importance of assessing management quality based on compliance to policy measures, strategic plan developments, risk taking aptitude and bank officers, managers and directors degree of involvement in decision-making.

Most importantly, regulatory parameters should extend to include all instruments, institutions and markets (see Avkiran et al. forthcoming; Moshirian 2011a). In cross border resolution, ring-fencing where bank assets are retained within borders was recommended (Makarova et al. 2010; Cerutti and Schmieder 2014). Policy makers should work further on the international financial architecture [see Moshirian (2002, 2011b)]. Moshirian (2011a) proposes to give large banks the liberty to grow. As such, this study recommends devising new policies that create convergence with soundness.

Acknowledgments Two gentlemen deserve more than this inadequate acknowledgement for their immense contribution to this chapter. A special debt of gratitude owed to Christian Ringle, for sparking off an interest in PLS-SEM, through his workshop in Leeds University, UK. Which has led me to begin my work and has culminated in this chapter. Christian has been extremely helpful throughout the write up of this paper, with his valuable suggestions and guidance. Without which this chapter would not have been possible. We also appreciate his patience and kindness to take time off to look through countless models. A special debt also extends far beyond to Necmi K. Avkiran for his invitation to contribute a sense of perspective for the finance discipline through this book chapter. Necmi has been extremely thorough in the editing of the chapter. He oversaw each phase of the writing consistently and tirelessly offering constructive feedback and seeing it through completion with a measure of compassion for which there can be no repayment.

References

Avkiran, N. K., Ringle, C. M., & Low, R. (forthcoming). Monitoring transmission of systemic risk: Application of PLS-SEM in financial stress testing. *The Journal of Risk.*

Bakir, C. (2013). *Bank behaviour and resilience: The effect of structures, institutions and agents.* Berlin: Springer.

Barr, R. S., & Siems, T. F. (1997). Bank failure prediction using DEA to measure management quality. In R. S. Barr, R. V. Helgason, & J. L. Kennington (Eds.), *Interfaces in computer science and operations research* (pp. 341–365). Boston: Springer.

Bastan, M., Mazrae, M. B., & Ahmadvand, A. (2016). Dynamics of banking soundness based on CAMELS rating system. In *The 34th International Conference of the System Dynamics Society. Delft, Netherlands.*

Bell, T. B. (1997). Neural nets or the logit model? A comparison of each model's ability to predict commercial bank failures. *Intelligent Systems in Accounting, Finance and Management, 6*(3), 249–264.

Berger, A. N., & Bouwman, C. H. (2009). Bank liquidity creation. *The Review of Financial Studies, 22*(9), 3779–3837.

Bernanke, B. (2007). Global imbalances: Recent developments and prospects. *Bundesbank Lecture Speech 4*, 18.

Bovenzi, J. F., Marino, J. A., & McFadden, F. E. (1987). Commercial bank failure prediction models. In D. J. Juttner & T. Valentine (Eds.), *The economics and management of financial institutions* (pp. 419–436). Melbourne: Longman Cheshire Pty.

Calomiris, C. W., & Herring, R. J. (2013). How to design a contingent convertible debt requirement that helps solve our too-big-to-fail problem. *Journal of Applied Corporate Finance, 25*(2), 39–62.

Canbas, S., Cabuk, A., & Kilic, S. B. (2005). Prediction of commercial bank failure via multivariate statistical analysis of financial structures: The Turkish case. *European Journal of Operational Research, 166*(2), 528–546.

Cerutti, E., & Schmieder, C. (2014). Ring fencing and consolidated banks' stress tests. *Journal of Financial Stability, 11*, 1–12.

Cheng, S. (2008). Board sensitivity and the variability of corporate performance. *Journal of Financial Economics, 87*(1), 157–176.

Chin, W. W. (1998). The partial least squares approach to structural equation modeling. *Modern Methods for Business Research, 295*(2), 295–336.

Costa Navajas, M., & Thegeya, A. (2013). Financial soundness indicators and banking crises.

Dang, U. (2011). *The CAMEL rating system in banking supervision: A case study of Arcada University of Applied Sciences.* International Business.

Davison, A. C., & Hinkley, D. V. (1997). *Bootstrap methods and their application*. Cambridge: Cambridge University Press.

Demirgüç-Kunt, A., & Detragiache, E. (2011). Basel core principles and bank soundness: Does compliance matter? *Journal of Financial Stability, 7*(4), 179–190.

Demirgüç-Kunt, A., Detragiache, E., & Tressel, T. (2008). Banking on the principles: Compliance with basel core principles and bank soundness. *Journal of Financial Intermediation, 17*(4), 511–542.

Diamantopoulos, A. (2006). The error term in formative measurement models: Interpretation and modeling implications. *Journal of Modelling in Management, 1*(1), 7–17.

Dijkstra, T. K. (2010). Latent variables and indices: Herman Wold's basic design and partial least squares. In V. V. Esposito, W. W. Chin, J. Henseler, & H. Wang (Eds.), *Handbook of partial least squares: Concepts, methods and applications* (Vol. 2, pp. 23–46). Berlin: Springer.

Dijkstra, T. K. (2014). PLS'Janus face–response to professor Rigdon's 'rethinking partial least squares modeling: In praise of simple methods'. *Long Range Planning, 47*(3), 146–153.

Edwards, J. R., & Bagozzi, R. P. (2000). On the nature and direction of relationships between constructs and measures. *Psychological Methods, 5*(2), 155–174.

Efron, B., & Tibshirani, R. J. (1993). *An introduction to the bootstrap*. New York: Chapman Hall.

Fama, E. F., & French, K. R. (1992). The cross-section of expected stock returns. *The Journal of Finance, 47*(2), 427–465.

Fornell, C., & Larcker, D. F. (1981). Evaluating structural equation models with unobservable variables and measurement error. *Journal of Marketing Research, 18*(1), 39–50.

Gaganis, C., Pasiouras, F., & Zopounidis, C. (2006). A multicriteria decision framework for measuring banks' soundness around the world. *Journal of Multi-Criteria Decision Analysis, 14*(1–3), 103–111.

Gasbarro, D., Sadguna, I. G. M., & Zumwalt, J. K. (2002). The changing relationship between CAMEL ratings and bank soundness during the Indonesian banking crisis. *Review of Quantitative Finance and Accounting, 19*(3), 247–260.

Gunsel, N. (2005). Financial ratios and the probabilistic prediction of bank failure in North Cyprus. *Editorial Advisory Board, 18*(2), 191–200.

Haenlein, M., & Kaplan, A. M. (2004). A beginner's guide to partial least squares analysis. *Understanding Statistics, 3*(4), 283–297.

Hair, J. F., Ringle, C. M., & Sarstedt, M. (2011). PLS-SEM: Indeed a silver bullet. *Journal of Marketing Theory and Practice, 19*(2), 139–152.

Hair, J. F., Jr., Hult, G. T. M., Ringle, C., & Sarstedt, M. (2017). *A primer on partial least squares structural equation modeling (PLS-SEM)*. Thousand Oaks: Sage.

Hanc, G. (1998). The banking crises of the 1980s and early 1990s: Summary and implications. *FDIC Banking Review, 11*, 1.

Henseler, J., Ringle, C. M., & Sinkovics, R. R. (2009). The use of partial least squares path modeling in international marketing. In R. R. Sinkovics & P. Ghauri (Eds.), *Advances in international marketing* (Vol. 20, pp. 277–319). Bingley: Emerald.

Ho, T., & Saunders, A. (1980). A catastrophe model of bank failure. *The Journal of Finance, 35*(5), 1189–1207.

Hooks, L. M. (1995). Bank asset risk: Evidence from early-warning models. *Contemporary Economic Policy, 13*(4), 36–50.

International Monetary Fund Staff. (2008). *Financial soundness indicators: Compilation guide*. Washington, DC: International Monetary Fund.

King, T. H. D., & Wen, M. M. (2011). Shareholder governance, bondholder governance, and managerial risk-taking. *Journal of Banking & Finance, 35*(3), 512–531.

Kodres, M. L. E., & Narain, A. (2010). *Redesigning the contours of the future financial system* (No. 2010). International Monetary Fund.

Korobow, L., & Stuhr, D. P. (1975). *Toward early warning of changes in bank's financial condition: A progress report*. na.

Korobow, L. E. O. N., Stuhr, D. P., & Martin, D. (1976). A probabilistic approach to early warning of changes in bank financial condition. *Monthly Review, 7*, 187–194.
Korobow, L., Stuhr, D. P., & Martin, D. (1977). A nationwide test of early warning research in banking. *Quarterly Review, 59*(3), 37–52.
Košak, M., Li, S., Lončarski, I., & Marinč, M. (2015). Quality of bank capital and bank lending behavior during the global financial crisis. *International Review of Financial Analysis, 37*, 168–183.
Kumar, M. A., Harsha, G. S., Anand, S., & Dhruva, N. R. (2012). Analyzing soundness in Indian banking: A CAMEL approach. *Research Journal of Management Sciences, 1*(3), 9–14.
Lee, L., Petter, S., Fayard, D., & Robinson, S. (2011). On the use of partial least squares path modeling in accounting research. *International Journal of Accounting Information Systems, 12*, 305–328.
Lindgren, C.J., Garcia, G., & Saal M. (1996). Bank soundness and macroeconomic policy. *IMF Occasional Papers*, No. 135, Washington, DC.
Loannidis, C., Pasiouras, F., & Zopounidis, C. (2010). Assessing bank soundness with classification techniques. *Omega, 38*(5), 345–357.
Lohmöller, J.-B. (1989). *Latent variable path modeling with partial least squares*. Heidelberg: Physica. doi:10.1007/978-3-642-52512-4
Makarova, M. Y., Ilyina, A., Schmieder, M. C., & Cerutti, E. (2010). *Bankers without borders? Implications of ring-fencing for European cross-border banks* (No. 10-247). International Monetary Fund.
Meyer, P. A., & Pifer, H. W. (1970). Prediction of bank failures. *The Journal of Finance, 25*(4), 853–868.
Moosa, I. A. (2010). Basel II as a casualty of the global financial crisis. *Journal of Banking Regulation, 11*(2), 95–114.
Moshirian, F. (2002). New international financial architecture. *Journal of Multinational Financial Management, 12*(4), 273–284.
Moshirian, F. (2011a). The global financial crisis and the evolution of markets, institutions and regulation. *Journal of Banking & Finance, 35*(3), 502–511.
Moshirian, F. (2011b). Global financial crisis, international financial architecture and regulation.
Nitzl, C. (2016). The use of partial least squares structural equation modelling (PLS-SEM) in management accounting research: Directions for future theory development. *Journal of Accounting Literature, 37*, 19–35.
Nurazi, R., & Evans, M. (2005). An Indonesian study of the use of CAMEL (S) ratios as predictors of bank failure. *Journal of Economic and Social Policy, 10*(1), 6.
Ohlson, J. A. (1980). Financial ratios and the probabilistic prediction of bankruptcy. *Journal of Accounting Research, 18*(1), 109–131.
Pettway, R. H., & Sinkey, J. F. (1980). Establishing on-site bank examination priorities: An early-warning system using accounting and market information. *The Journal of Finance, 35*(1), 137–150.
Podviezko, A., & Ginevičius, R. (2010). *Economic criteria characterising bank soundness and stability*. 6th International Scientific Conference, May 13–14, 2010, Vilnius, Lithuania. Business and Management 2010.
Poon, W. P., Firth, M., & Fung, H. G. (1999). A multivariate analysis of the determinants of Moody's bank financial strength ratings. *Journal of International Financial Markets, Institutions and Money, 9*(3), 267–283.
Rasli, A., Goh, C. F., & Khan, S. U. R. (2013). Demystifying the role of a state ownership in corporate governance and firm performance: Evidence from the manufacturing sector in Malaysia. *Zbornik radova Ekonomskog fakulteta u Rijeci: časopis za ekonomsku teoriju i praksu, 31*(2), 233–252.
Ratnovski, L., & Huang, R. (2009). Why are Canadian banks more resilient? *IMF Working Paper*, 152(9).

Reinartz, W., Haenlein, M., & Henseler, J. (2009). An empirical comparison of the efficacy of covariance-based and variance-based SEM. *International Journal of Research in Marketing, 26*(4), 332–344.
Richter, N. F., Cepeda, G., Roldán, J. L., & Ringle, C. M. (2015). European management research using partial least squares structural equation modeling (PLS-SEM). *European Management Journal, 33*(1), 1–3.
Rigdon, E. E. (2012). Rethinking partial least squares path modeling: In praise of simple methods. *Long Range Planning, 45*(5), 341–358.
Rigdon, E. E. (2014). Rethinking partial least squares path modeling: Breaking chains and forging ahead. *Long Range Planning, 47*(3), 161–167.
Roman, A., & Şargu, A. C. (2013). Analysing the financial soundness of the commercial banks in Romania: An approach based on the camels framework. *Procedia Economics and Finance, 6*, 703–712.
Sarker, A. (2006). CAMELS rating system in the context of Islamic banking: A proposed 'S'for Shariah framework. *Journal of Islamic Economics, Banking and Finance, 2*(2), 1–26.
Schaeck, K., & Cihak, M. (2012). Banking competition and capital ratios. *European Financial Management, 18*(5), 836–866.
Sinkey, J. F. (1979). *Problem and failed institutions in the commercial banking industry* (pp. 34–39). Greenwich: Jai Press.
Tenenhaus, M., & Vinzi, V. E. (2005). PLS regression, PLS path modeling and generalized procrustean analysis: A combined approach for multiblock analysis. *Journal of Chemometrics, 19*(3), 145–153.
Toor, N. S. (2006). *Handbook of banking information*. New Delhi: Skylark.
Tussing, A. D. (1967). The case for bank failure. *The Journal of Law and Economics, 10*, 129–147.
Währungsfonds, I. (2009). *Global financial stability report–responding to the financial crisis and measuring systemic risk*. April, Washington, DC.
Wold, H. (1982). Models for Knowledge. *The Making of Statisticians*, 189–212. doi:10.1007/978-1-4613-8171-6_14.
Wold, H. (1985). Partial least squares. In S. Kotz & N. L. Johnson (Eds.), *Encyclopedia of statistical sciences* (Vol. 6, pp. 581–591). New York: Wiley.

Chapter 3
The Customer Loyalty Cascade and Its Impact on Profitability in Financial Services

Anne-Kathrin Hegner-Kakar, Nicole F. Richter, and Christian M. Ringle

Abstract Building and maintaining successful, long-term relationships is one of the crucial tasks in today's financial sector, the idea being that loyal customers purchase more, demonstrate a higher willingness to spend, and act as advocates for the company. However, there is also some controversy on whether profitability indeed increases with customer loyalty. We will analyze the process of loyalty development and evaluate if and how customer loyalty (positively) affects profitability. We will do so with reference to a four stage sequential loyalty model that grounds on a chain of effects from cognitive loyalty, affective loyalty, conative loyalty to action loyalty. We will make use of PLS structural equation modeling and analyze data of almost 7000 customers of a German bank surveyed by telephone. These analyses will support practitioners in the banking and financial sector in setting-up and steering their customer retention strategies and will provide a theoretical contribution to validating one of the most prominent customer loyalty models.

Keywords Customer loyalty · Profitability · Financial services · Partial least squares · Path model · PLS-SEM

A.-K. Hegner-Kakar (✉)
Fashion and Lifestyle, GfK SE, Nürnberg, Germany

Hamburg University of Technology (TUHH), Hamburg, Germany
e-mail: anne-kathrin.hegner-kakar@gfk.com

N.F. Richter
University of Southern Denmark, Sønderborg, Denmark
e-mail: nicole@sam.sdu.dk

C.M. Ringle
Institute of HRM, Hamburg University of Technology, Hamburg, Germany

The University of Newcastle, Faculty of Business and Law Callaghan, Callaghan, NSW, Australia
e-mail: c.ringle@tuhh.de

N.K. Avkiran, C.M. Ringle (eds.), *Partial Least Squares Structural Equation Modeling*, International Series in Operations Research & Management Science 267, https://doi.org/10.1007/978-3-319-71691-6_3

3.1 Introduction

Building and maintaining successful, long-term relationships is crucial in many of today's competitive environments, in particular in the financial services sector. The costs for acquiring new customers is five times higher than maintaining existing customers. For credit cards, reducing customer defections by 5% doubles the average customer life span, and increases profit by 75% (Athanasopoulou 2009; Reichheld and Sasser 1990). Losing customers is costly, as loyal customers purchase more, demonstrate a higher willingness to spend, and act as advocates for the company (Harris and Goode 2004). These arguments are of special relevance in the financial services sector as it faces the challenge of partial customer defection more than other industries. Although only a minority of customers actually leaves their financial service provider, many customers simultaneously enter relationships with other providers. Specifically, bank customers spread their portfolio of services and products to different institutions, thereby reducing the individual firm's share of wallet (Aurier and N'Goala 2010; Du et al. 2007).

Given the acceptance of the link between loyalty and financial outcomes (Anderson and Mansi 2009; Anderson and Mittal 2000; Edvardsson et al. 2000; Fornell et al. 2006, 2009; Hallowell 1996; Lariviere 2008; Rust and Zahorik 1993), long-term relationships and the development of customer loyalty have become a central issue in marketing (Athanasopoulou 2009). However, there is also some controversy on whether profitability indeed increases with customer loyalty. For instance, Reinartz and Kumar (2000) show in a non-contractual setting that both short- and long-term relationships can be highly profitable. Their findings strongly challenge the notion that long-life customers trigger increased lifetime spending, lower costs of servicing, and reduced price sensitivity (Reinartz and Kumar 2000). Not all long-term customers are necessarily profitable, so focusing on the most profitable customer segments becomes crucial for companies (Niraj et al. 2001; Storbacka et al. 1994). As a result, the managerial focus shifted from "(…) mere 'loyalty' to 'customer profitability' and 'profitable loyalty'"(Kumar 2016, p. 108).

We will analyze the process of loyalty development and evaluate if and how customer loyalty (positively) affects profitability. We will do so with reference to a four stage sequential loyalty model that grounds on a chain of effects from cognitive loyalty, affective loyalty, conative loyalty to action loyalty. This analysis enables us to answer some of the key questions in today's research landscape: How can we describe the chain of effects between different stages of loyalty development? Which loyalty stages exert an impact on profitability? How can companies leverage this knowledge to drive loyalty and profit, or simply profitable loyalty? We will make use of a sample of almost 7000 customers of a German bank surveyed by telephone. Specifically, we will apply PLS structural equation modeling (PLS-SEM; Hair et al. 2017a, 2018, in press) to test the effect of the four sequential customer loyalty stages on profit. These analyses will support practitioners in the banking and financial sector in setting-up and steering their customer retention strategies and will provide a theoretical contribution to validating one of the most prominent customer loyalty models (Oliver 1999).

3.2 The Dynamics of Customer Relationships: A Sequential Model of Profitable Loyalty

3.2.1 Theoretical Perspectives on Relational Exchanges

"Marketing is to establish, maintain, and enhance relationships with customers and other partners, at a profit, so that the objectives of the parties involved are met" (Grönroos 1990, p. 138). To explain these relationships and the mechanisms through which they become loyal and profitable, several different theoretical paradigms are referred to, with Transaction Cost Theory (Williamson 1975, 1985), Relational Contracting Theory (Macneil 1978), Social Exchange Theory (Blau 1964; Thibaut and Kelley 1959), Resource Dependence Theory (Pfeffer and Salanzik 1978), and the Resource-Based View (Dyer 1996; Jap 1999, 2001) representing the most prominent frameworks. Each of these theoretical perspectives focusses on different drivers of exchange performance (Eiriz and Wilson 2006; Palmatier et al. 2006).

Transaction Cost Theory views value in a relationship as a purely rational evaluation of expected financial benefits and gains (Gassenheimer et al. 1998; Williamson 1975) and puts its primary emphasis on single transactions lacking a dynamic perspective (Doz and Prahalad 1991; Joshi and Stump 1999). It is therefore not the first choice theory for our purposes, as we are researching into an ongoing service contract. Similarly, Relational Exchange Theory is less suitable for our purposes due to its focus on relational norms. As Resource Dependence Theory and the Resource-Based View mainly concentrate on inter-organizational relationships, these frameworks are not considered first choice for a business-to-consumer context.

Instead, researchers analyzing relational exchange have mainly drawn on **Social Exchange Theory** (SET) which concentrates on the relational contract that develops over time (Lambe et al. 2000). Grounded on works in interpersonal relationships and sociology (Blau 1964; Thibaut and Kelley 1959), SET emphasizes the processes that lead to satisfaction for the parties involved in the exchange (Cannon and Perreault 1999). It focuses on the social benefits a relationship entails, placing emphasis on the satisfaction with the cooperation and a comparative evaluation of alternative partners (Thibaut and Kelley 1959). Accordingly, a relational exchange holds a social and economic dimension, as both the exchange process itself as well as the utility rendered are important (Singh and Sirdeshmukh 2000). SET explicitly considers exchange relationships as dynamic processes which evolve over time as the exchange partners interact and demonstrate their trustworthiness and commitment (Hallén et al. 1991). SET has also influenced popular theories such as the "commitment-trust theory" by Morgan and Hunt (1994) which identifies commitment and trust as key variables in relationships. Typical antecedents of relationship performance in this context include relationship quality, relational benefits, value, commitment, trust, and satisfaction (Chaudhuri and Holbrook 2001; Garbarino and Johnson 1999; Harris and Goode 2004; Hennig-Thurau et al. 2002; Morgan and Hunt 1994; Papassapa and Miller 2007).

We will draw on SET to explain the four stage loyalty model. Before presenting the model and research hypotheses, the concept of customer loyalty is briefly introduced.

3.2.2 *The Customer Loyalty Cascade*

All relationship marketing activities are eventually evaluated based on profit, customer loyalty and word-of-mouth communications (Hennig-Thurau et al. 2002). The concept of customer loyalty, until the 1970s, was considered solely as repeat purchasing behavior, induced mainly by repeated purchase cycles that were stochastic in character (Evanschitzky and Wunderlich 2006). Jacoby and Kyner (1973) were among the first scholars to introduce a new behavioral perspective, based on a biased (i.e., non-random) repeat purchase and a deliberate selection of a specific brand among a set of alternatives that results from an evaluation process. Day (1969) suggested that focusing on purchase behavior as behavioral loyalty will confound true loyalty and spurious loyalty and recommended combining attitudinal and behavioral loyalty. Extending the mere behavioral definition, Jacoby and Chestnut (1978) and Dick and Basu (1994) made seminal contributions to establish the multi-dimensional nature of loyalty. Today, scholars have adopted this multi-dimensional view of loyalty and agree that loyalty includes both behavioral and attitudinal aspects (Brunner et al. 2008; Chaudhuri and Holbrook 2001; Chiou and Droge 2006; Papassapa and Miller 2007).

Building on this multidimensional view of customer loyalty, Oliver (1997, 1999) developed a detailed framework of a four-stage sequential loyalty model based on a chain of effects from cognitive loyalty, affective loyalty, conative loyalty to action loyalty (see Fig. 3.1) which forms the basis of many empirical studies testing for loyalty determinants (e.g., Ahrholdt et al. 2016). Accordingly, customers become loyal in a cognitive sense first, then subsequently in an affective, conative and finally in a behavioral manner. Based on the assumption that the four distinct dimensions of customer loyalty do not appear simultaneously, Oliver (1999) suggested they emerge consecutively over time and that different factors influence customer loyalty at each stage. In this framework, customer loyalty is only achieved if high loyalty is attained across all four loyalty stages of the "customer loyalty cascade" (Han et al. 2008). This notion of a sequential loyalty chain is in line with frameworks such as the satisfaction-profit chain (Anderson and Mittal 2000), which conceptualize the chain of

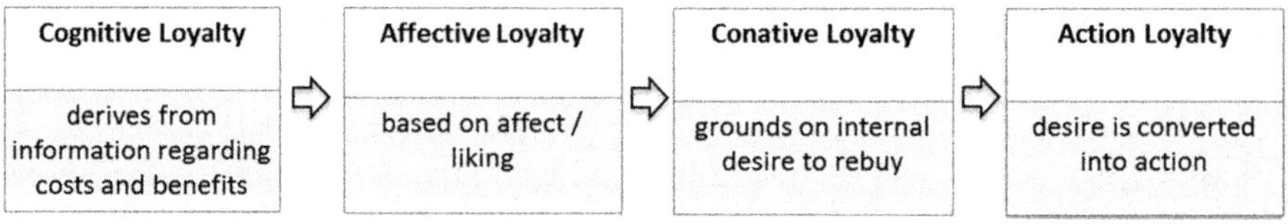

Fig. 3.1 Four-stage loyalty model. Source: Adapted from Oliver (1999)

effects in which customer loyalty mediates the effect of customer satisfaction to profitability. A few empirical studies have confirmed Oliver's (1999) sequential loyalty chain in do-it-yourself retailing (Evanschitzky and Wunderlich 2006), services (Han et al. 2008), online services (Harris and Goode 2004), seafood products (Olsen 2002), and department stores (Sivadas and Baker-Prewitt 2000).

The first loyalty phase is **cognitive loyalty** which derives from information about a company's offering, such as quality and price. This information indicates that the selected product or service is the best choice among its alternatives and thus preferable to others. Oliver (1999) argues that cognitive loyalty constitutes the weakest type of loyalty because it does not relate to the brand. Instead, it derives from information about costs and benefits of a company's offering and is therefore also referred to as "phantom loyalty" (Oliver 1999, p. 37). Switching behavior may occur in case of superior competitive offerings. But when this information is processed, it becomes part of the consumer's experience and can entail an affective reaction derived from cumulatively satisfying experiences (Oliver 1999). Service quality refers to such a cognitive state, while satisfaction is the affective (or emotional) state resulting from an evaluation of interaction experiences (Carrillat et al. 2009; Crosby et al. 1990).

Affective loyalty constitutes the second phase of loyalty and refers to a favorable attitude (or liking) towards the company or brand and comprises a sense of pleasurable fulfillment. The social exchange approach emphasizes the belief in mutually beneficial behavior and states that only social exchange can entail feelings of gratitude or trust, a mere economic exchange cannot (Blau 1964; Seppänen et al. 2007). Social relationships thus represent a source of positive or negative emotions (Lawler 2001). As a service encounter reflects on the customer's view of the entire company, positive (negative) service encounters will lead to positive (negative) feelings towards the service provider (Sierra and McQuitty 2005). Every experience with a firm (or the cognitive evaluation of it) can influence affective responses such as cumulative satisfaction and trust (Brunner et al. 2008, Chiou and Droge 2006; Garbarino and Johnson 1999; Oliver 1997). Satisfaction acts as an anchor that is updated with the information gained in every consumption experience (Bolton 1998). Hence, we hypothesize:

Hypothesis 1 (H1): Cognitive loyalty has a positive direct effect on affective loyalty.

According to SET, both economic rewards (e.g., value) and social rewards (e.g., trust) represent important determinants of the exchange partner's behavior (Emerson 1976). The higher the level of satisfaction with the rewards, the more likely will the exchange partners continue the relationship (Briggs and Grisaffe 2009). In other words, with repeatedly satisfying experiences, strengthened beliefs, and affect, the customer may develop a motivation to rebuy the brand (Oliver 1999). This motivation is described in the third phase of loyalty.

Conative loyalty, the most studied dimension of customer loyalty (Han et al. 2008): Conative loyalty describes an internal desire to repurchase a certain brand which binds the customer more strongly to the company than affective loyalty

(Brunner et al. 2008; Oliver 1999) and is characterized by a deeper level of commitment (Harris and Goode 2004). As emotions will impact the customer's loyalty towards the company (Sierra and McQuitty 2005), satisfaction leads to repurchase intentions (Cronin and Taylor 1992; Papassapa and Miller 2007). We expect:

Hypothesis 2 (H2): Affective loyalty has a positive direct effect on conative loyalty.

As with any intention, conative loyalty represents an internal disposition to rebuy, but this motivation may finally not be translated into real behavior (Brunner et al. 2008; Oliver 1999). Action loyalty describes the actual behavior in which the preposition or readiness to repurchase a firm's offering developed in the previous loyalty stages is converted into action. Behavioral intentions elicit a behavioral response, including repurchase or customer retention (Bolton 1998; Perkins-Munn et al. 2005; Rust and Zahorik 1993) and increased usage (Bolton and Lemon 1999). Vulnerabilities of this conversion into action may elicit from consumer idiosyncrasies (such as variety seeking) or switching incentives (stemming for instance from the competitive environment). Nonetheless, we hypothesize:

Hypothesis 3 (H3): Conative loyalty has a positive direct effect on action loyalty.

Summarizing the above considerations, we agree with Oliver's (1999) conceptualization and previous research (Evanschitzky and Wunderlich 2006; Han et al. 2008; Harris and Goode 2004; Sivadas and Baker-Prewitt 2000) and expect a direct positive relationship from each loyalty phase to the subsequent stage.

However, some authors highlight that customer loyalty may be more complex than depicted by the linear, sequential structure of the four stage loyalty model (Agustin and Singh 2005; Han et al. 2008). There are also theoretical considerations supporting additional relationships in the model based on SET. When assessing the satisfaction with a relationship, customers evaluate the relationship against a comparison standard of alternatives. If the outcomes in a relationship fall below this level and there are viable alternatives, the partner will terminate the relationship. Hence, this evaluation influences the decision whether to continue or leave a relationship (Thibaut and Kelley 1959). So a customer may not be satisfied with a company but still remain in the relationship due to a lack of superior alternatives. This may indicate a direct relationship between this comparative evaluation of alternatives (cognitive loyalty) as well as the intention to stay loyal (conative loyalty) and actual behavior (action loyalty). Therefore, we also test whether the relationships between loyalty dimensions on three different stages are fully mediated by the construct in between or whether there are also direct relationships surpassing the mediating construct. We will outline the following hypotheses building on the model outlined by Oliver (1999):

Hypothesis 4 (H4): The effect from cognitive loyalty to conative loyalty is fully mediated by affective loyalty.

Hypothesis 5 (H5): The effect from cognitive loyalty to action loyalty is fully mediated by affective loyalty and conative loyalty.

Hypothesis 6 (H6): The effect from affective loyalty to action loyalty is fully mediated by conative loyalty.

3.2.3 *The Impact of Loyalty on Profitability*

Although Oliver's (1999) sequential loyalty model has been confirmed in earlier studies (Evanschitzky and Wunderlich 2006; Han et al. 2008; Harris and Goode 2004; Sivadas and Baker-Prewitt 2000), the (relative) impact of each of the four individual loyalty stages on financial outcomes has—to the best of our knowledge—not been tested so far. Measuring the effect of loyalty on profitability is essential as "all relationship marketing activities are ultimately evaluated on the basis of the company's overall profitability" (Hennig-Thurau et al. 2002, p. 231). In general, we follow the logic that profits occur when a customer has been loyal for some time, as costs tend to be generated in the beginning of a relationship. The more loyal a customer and the longer the customer stays in the relationship, the larger the sales and profits (Edvardsson et al. 2000).

Some studies confirm a positive effect of action loyalty on customer profitability (Helgesen 2006). Similarly, there is evidence for a positive relationship between trust and satisfaction (i.e., affective loyalty) on financial outcomes (Anderson et al. 1994; Edvardsson et al. 2000; Hallowell 1996; O'Sullivan and McCallig 2009; Palmatier et al. 2006). Satisfaction may exert a direct effect on profit because it accounts for those contributions to revenues or cost savings that are based on a positive experience. Specifically, operating costs only decrease if the customer is satisfied as costs associated with service guarantees or complaint handling will not arise. Satisfied customers should be more willing to pay for the benefits they get and tend to be more willing to accept price increases. Conversely, the indirect effect via loyalty stems from revenue growth and higher profitability resulting from repeat purchases, cross-selling, and higher willingness to pay more (Anderson et al. 1994; Edvardsson et al. 2000). However, there is evidence that relational effects tend to be more closely related to loyalty dimensions in the cascade than to objective firm performance, for which other factors outside the relationship might be more relevant (e.g., the competitive environment or economic situation) (Palmatier et al. 2006). To contribute to a better understanding of the differential effects of loyalty phases on firm outcomes, this study will also pay special attention to determining the impact of each of the four loyalty stages on customer profitability. Hence, we outline the following hypotheses related to affective and action loyalty building on former research, yet also to cognitive and conative loyalty to be tested more exploratively:

Hypothesis 7 (H7): Cognitive loyalty has a positive direct effect on profitability.

Hypothesis 8 (H8): Affective loyalty has a positive direct effect on profitability.

Hypothesis 9 (H9): Conative loyalty has a positive direct effect on profitability.

Hypothesis 10 (H10): Action loyalty has a positive direct effect on profitability.

3.3 Research Design

3.3.1 Sample and Data Collection

To test our hypotheses, we draw on a sample from the financial industry. The financial sector faces high costs of acquiring new customers, low customer loyalty and a rather high number of unprofitable customers (Di Benedetto and Kim 2016; Mittal and Katrichis 2000). A deeper understanding of the dynamics of customer relationships, therefore, is of specific practical relevance to firms and institutions in the industry.

We conducted telephone interviews with more than 6852 (German) customers of one bank headquartered in Germany in 2010. In addition to the primary data collected by means of the survey, we gathered secondary data on the customers, namely figures on customer profitability received from the management accounting of the bank.

3.3.2 Research Model and Analysis Technique

To test our hypotheses, we make us of PLS-SEM and SmartPLS 3.0 (Ringle et al. 2015). We opted for using PLS-SEM (e.g., Hair et al. 2017a, 2018, in press) as a tool that is very useful for developing and extending existing theory in management research. PLS-SEM has benefits for prediction, has the power to easily cope with formative constructs involved, and has strong features for testing mediating effects and complex models (e.g., Hair et al. 2017c; Richter et al. 2016a, 2016b; Sarstedt et al. 2017).

Our analysis involves two steps (see Fig. 3.2). First, we focus on testing the validity of the four-stage loyalty model, namely our hypotheses H1–H6. We do so by performing mediator-analyses which test whether the indirect and/or direct effects between the four constructs are significant or not. Following the procedures outlined in Zhao et al. (2010) and Nitzl et al. (2016), we determine the type of mediation between the constructs. More specifically, we test whether there is a full mediation of the constructs as defined by Oliver (1999). If this assumption holds true, we would expect the paths d, e, and f in Fig. 3.2 to be not significant, yet the indirect effects (of cognitive on conative via affective loyalty, and of cognitive on

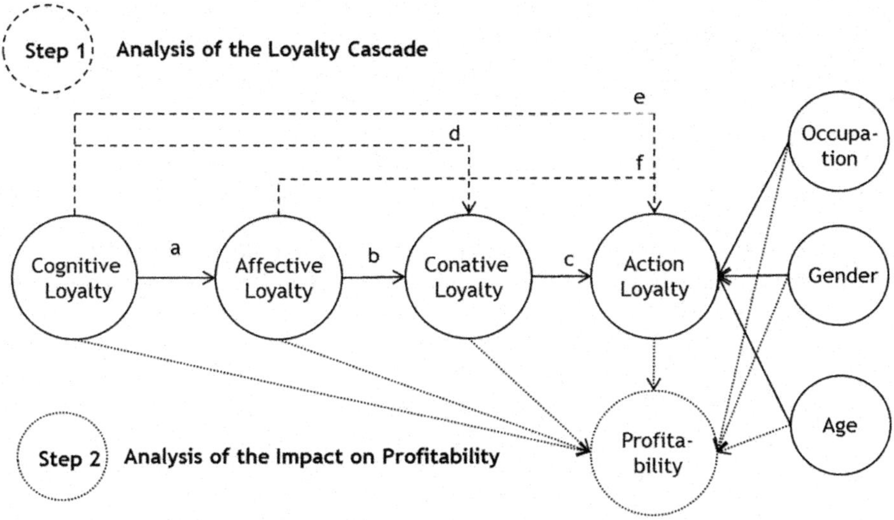

Fig. 3.2 A two-step research approach

action via affective and conative loyalty, and of affective on action loyalty via conative loyalty) to be significant (Hair et al. 2017b).

Second, we test the effect of each of the four loyalty stages on customer profitability, i.e., our Hypotheses 7–10. Thus, we will close the research gap which exists on individual customer level; research mostly focused on the effects of loyalty on firm performance at a company level (Edvardsson et al. 2000), and only a few studies looked at it from an individual customer level (e.g., Helgesen 2006). In both analyses, we control for the effects of gender, age and occupation on action loyalty and on customer profitability.

3.3.3 Measures

Researchers can generally choose between two different types of measurement models (e.g., Coltman et al. 2008; Diamantopoulos and Winklhofer 2001; Gudergan et al. 2008): reflective measurement models and formative measurement models. Reflective measurement models have direct relationships from the construct to the indicators and treat the indicators as error-prone manifestations of the underlying construct. In contrast, in a formative measurement model, a linear combination of a set of indicators forms the construct (i.e., the relationship is from the indicators to the construct). Hence, "variation in the indicators precedes variation in the latent variable" (Borsboom et al. 2003, p. 208). Indicators of formatively measured constructs do not necessarily have to correlate strongly as is the case with reflective indicators. In PLS-SEM, formative measurement models represent composite indicator models (Sarstedt et al. 2016b, 2017). Finally, researchers often use single-item constructs. However, if not included as a control or a moderator variable, single-item constructs usually are disadvantageous compared to multi-item constructs (Diamantopoulos et al. 2012; Sarstedt et al. 2016a).

The measurement theory specifies how to measure latent variables. Thereby, researchers define the type of measurement model and the relevant indicators. In the following we define the measures based on the findings of prior research:

- We use a single-item measure to operationalize **customer profitability** on the individual customer level. Building on the definition of Storbacka et al. (1994), we measure customer profitability as the *total revenues* generated from a customer relationship *minus the total costs* from serving a customer relationship over a fiscal year (e.g., Lariviere 2008).
- **Action loyalty** is operationalised by means of two (formative) indicator variables. Following former operationalisations (Hallowell 1996), it is measured based on the *relative length of the customer relationship* and the *cross-buying intensity* per customer.
- We use two (reflective) items to operationalize **conative loyalty** which are in line with past operationalisations (Brunner et al. 2008; Evanschitzky and Wunderlich 2006): *"Based on your experience, would you select [BANK] as your bank again?"*, and *"How likely is it that you would recommend [BANK] to a friend or relative?"*.
- We use the following five (reflective) items to measure **affective loyalty** based on an affective reaction through overall satisfaction and trust (Carrillat et al. 2009; Delgado-Ballester and Munuera-Alemán 2001). The items all reflect aspects of affective loyalty related to overall satisfaction, honesty and benevolence (Evanschitzky and Wunderlich 2006; Kantsperger and Kunz 2010). Overall satisfaction is reflected in the item *"Overall, how satisfied are you with [BANK]?* "(Hallowell 1996; Sivadas and Baker-Prewitt 2000)). *"[BANK] is a bank, I have full trust in"* and *"[BANK] is honest and always keeps its promises"* are two items adapted from Doney and Cannon (1997) and Dagger and O'Brien (2010) to cover the honesty aspects of affective loyalty. Finally, we refer to *"[BANK] respects me as a customer"* and *"[BANK] always treats me seriously"* to measure the benevolence aspects of affective loyalty, building on Sirdeshmukh et al. (2002).
- We use 7 (formative) items to operationalize **cognitive loyalty**. Two items cover the two defining dimensions of calculative commitment, namely the evaluation of a relationship's costs and benefits, and the lack of better alternatives (Hansen et al. 2003): *"[BANK] offers the best value for money"* and *"[BANK] is always one step ahead of other banks"*. Furthermore, five items relate to the dimensions of the SERVQUAL instrument: *tangibles*, *reliability, responsiveness, assurance* and *empathy* (Parasuraman et al. 1985, 1988; Rust and Zahorik 1993). Although other measures of service quality were also successfully applied (e.g., Avkiran 1994, 2002), using (the performance assessment involved in) SERVQUAL in financial services is common and proved to be useful in former studies (e.g., Arasli et al. 2005; Cronin and Taylor 1992). The five items

finally in use were generated by means of five individual factor analyses on individual items as found in previous studies (e.g., Mukherjee and Nath 2005) before importing them into SmartPLS. For these factor analyses, the loadings were all on appropriate levels (with 0.766 as the lowest factor loading), as were the shares of explained variances of the factors (with 69% as the lowest share of explained variance found for assurance).

For further information on the individual items, refer to Tables 3.1 and 3.2 which provide an overview of all items used including quality criteria. Finally, we include the following single-items as **control variables** into the analysis: gender, age and occupation of the customer.

3.4 Discussion of Results

3.4.1 Evaluation of Measurement Models

Before interpreting the results of our structural models, we evaluate the quality of our measurement models (e.g., Chin 2010; Sarstedt et al. 2017; Tenenhaus et al. 2005). As summarized in Table 3.1, all of our reflective measurement models perform well with regard to standard quality criteria: All outer loadings are above 0.70; even the item with the lowest loading (*"Overall, how satisfied are you with [BANK]?"*) still is (with 0.752) well above the threshold value. The average variances extracted of affective loyalty (0.699) and conative loyalty (0.820) exceed 0.50 indicating appropriate convergent validity of the constructs. Finally, the composite reliabilities are above 0.90, yet below 0.95 for both reflective constructs (with 0.921 for affective loyalty and 0.901 for conative loyalty), which indicates appropriate construct reliability (Hair et al. 2017b). The evaluation of collinearity along the variance inflation factors does not point to collinearity issues for our formative indicators (see Table 3.2). Most of our formative weights are moreover significant, yet not all. For the indicator variables showing insignificant weights, we opted for not eliminating the variables from further analysis after evaluating loadings, significance of loadings and reassessing the theoretical contents of the constructs to be covered.

3.4.2 Step 1: Analysis of the Loyalty Cascade

Table 3.3 presents the results of our mediator analysis. It provides the path coefficients for all direct and indirect effects including information on their significance based on bias-corrected and accelerated (BCa) bootstrap confidence intervals. Furthermore, variance inflation factors are given. The variance inflation factors do not point to problems of collinearity in the structural model. Furthermore the model shows a high share of explained variance for affective (R^2 = 0.619) and conative

Table 3.1 Reflective and single-item measurements and quality criteria

Reflective and single-item measures	Loading >0.70	AVE >0.50	CR >0.70	HTMT CI without 1
Customer profitability				
Total revenues—total costs of serving the customer, categorized into four groups: negative, low, medium and high	1			
Conative loyalty				
Based on your experience, would you select [BANK] as your bank again? (1 = no, definitely not, 5 = yes, definitely)	0.911	0.820	0.901	[0.892; 0.909]
How likely is it that you would recommend [BANK] to a friend or relative? (0 = surely not, 10 = yes, definitely)	0.900			
Affective loyalty				
Overall, how satisfied are you with [BANK]? (1 = dissatisfied, 5 = very satisfied)	0.752	0.699	0.921	[0.916; 0.925]
[BANK] is a bank that I have full trust in (1 = I do not agree at all, 7 = I fully agree)	0.866			
[BANK] is honest and always keeps its promises. (1 = I do not agree at all, 7 = I fully agree)	0.845			
[BANK] respects me as a customer (1 = I do not agree at all, 7 = I fully agree)	0.850			
[BANK] always treats me seriously (1 = I do not agree at all, 7 = I fully agree)	0.863			
Controls				
Gender (1 = male, 0 = female)	1			
Age	1			
Occupation (1 = blue collar, 0 = white collar)	1			

Note: *AVE* average variance extracted, *CR* composite reliability, *CI* 95% bias corrected confidence interval, *HTMT* heterotrait-monotrait-ratio

loyalty ($R^2 = 0.569$). However, the model shows a very poor share of explained variance for action loyalty ($R^2 = 0.040$).

The four stage model indicates that affective loyalty is positively and directly impacted by cognitive loyalty (H1). Indeed the relationship between the two constructs shows the highest and a significant path coefficient (0.787). Second, the model indicates that conative loyalty is positively and directly impacted by affective loyalty (H2). This also receives strong support with a significant path coefficient of 0.571. Third, the model suggests that action loyalty is positively and directly impacted by conative loyalty (H3). This is not supported; our results do not show a significant path coefficient for this relationship.

Furthermore, the following mediations are implicit in the four stages of the model: First, cognitive loyalty impacts conative loyalty via affective loyalty, i.e., the relationship between the two is fully mediated by affective loyalty (H4). For this to hold true, the indirect effect of cognitive and conative loyalty via affective loyalty ($a \times b$) needs to be significant while the direct effect (d) is not significant. The indi-

Table 3.2 Formative measurements and quality criteria

Formative measures	Weights	P	P < 0.10	VIF < 5
Action loyalty				
Length of the customer relationship in months related to age	0.133	0.471	No	1.002
Cross-buying intensity (number of products owned from the bank)	0.986	0.000	Yes	1.002
Cognitive loyalty				
[BANK] offers the best value for money. (1 = I do not agree at all, 7 = I fully agree)	0.360	0.000	Yes	1.649
[BANK] is always one step ahead of other banks. (1 = I do not agree at all, 7 = I fully agree)	0.542	0.000	Yes	1.711
Tangibles (factor score) from two items: (1) location and accessibility of the branch, (2) cleanliness and order in the branch	−0.030	0.197	No	1.963
Reliability (factor score) from two items: (1) banking personal can be reached easily, (2) reliability of statements made by the consultant	0.298	0.000	Yes	3.447
Responsiveness (factor score) from two items: (1) Motivation to help, (2) Length of the waiting time before you speak to a consultant	0.056	0.039	Yes	3.097
Assurance (factor score) from three items: (1) competency of the consultant, (2) friendliness of the consultant, (3) discretion in the branch	0.032	0.271	No	3.616
Empathy (factor score) from two items: (1) opening hours of the branch, (2) time the consultant took for customer	−0.010	0.652	No	3.159

Note: *VIF* variance inflation factor

rect effect is with a value of 0.449 highly significant. The direct effect of cognitive on conative loyalty is, however, likewise quite strong (0.218) and significant. Hence, there is only a partial mediation, more specifically there is a complementary partial mediation. This indicates that a portion of the effect of cognitive loyalty on conative loyalty is mediated through affective loyalty, yet cognitive loyalty still explains a portion of conative loyalty that is independent of affective loyalty. This partly contradicts our hypothesis H4.

Second, the model suggests that cognitive loyalty impacts action loyalty via affective and conative loyalty, i.e., the relationship between the two is fully mediated by affective and conative loyalty (H5). The indirect effect (a × b × c) is with a value of 0.041 not significant; the direct effect of cognitive loyalty on action loyalty (e) is likewise not significant. Hence, the results do neither support a mediated effect nor a direct effect of cognitive on action loyalty, and thus, contradict our hypothesis H5.

Third, the model implies that affective loyalty impacts action loyalty via conative loyalty, i.e., the relationship between the two is fully mediated by conative loyalty (H6). The indirect effect of affective loyalty on action loyalty (b × c) is not significant; the direct effect (f) likewise is not significant. Hence, the results do neither support a

Table 3.3 Results of the mediator analyses of the loyalty cascade (Step 1)

Relationship	Path coefficient [95% confidence interval]		VIF	Significant? [building on confidence intervals]
Indirect effects significant?				
Cognitive ➔ conative loyalty (via affective loyalty) (a × b)	0.449 [0.412;0.480]	***		Yes
Cognitive ➔ action loyalty (via affective & conative loyalty) (a × b × c)	0.041 [−0.016;0.109]			No
Affective ➔ action loyalty (via conative loyalty) (b × c)	0.022 [−0.005; 0.048]			No
Direct effects significant?				
(a) Cognitive ➔ affective loyalty	0.787 [0.770;0.801]	***	1.000	Yes
(b) Affective loyalty ➔ conative loyalty	0.571 [0.526; 0.610]	***	2.623	Yes
(c) Conative loyalty ➔ action loyalty	0.035 [−0.010; 0.086]		2.321	No
(d) Cognitive ➔ conative loyalty	0.218 [0.176; 0.259]	***	2.623	Yes
(e) Cognitive ➔ action loyalty	−0.054 [−0.106; 0.002]		2.734	No
(f) Affective ➔ action loyalty	0.022 [−0.060; 0.093]		3.379	No

Note: ***$p < 0.001$; *VIF* variance inflation factor

mediated effect nor a direct effect of affective loyalty on action loyalty. Thus, results contradict our hypothesis H6.

3.4.3 Step 2: Impact on Profitability

In the second step we test for the effects of the loyalty dimensions on customer profitability. Figure 3.3 presents the results of the model involving the four loyalty dimensions along the four stage approach as well as the direct effect of cognitive on conative loyalty which showed significance in our mediator analyses. It provides the path coefficients, their significance and R-square values.

The model explains a very low share of variance in customer profitability, namely around 2% ($R^2 = 0.02$). The only loyalty construct that has a significant effect on customer profitability is action loyalty (with a path of 0.14). All other loyalty constructs do not directly impact customer profitability. In addition to the path coefficients given in the figure, we calculated effect sizes for the relationships which enables further testing of the strength of the effect of action loyalty on profitability. The f^2 effect size of action loyalty on customer profitability is 0.018. Following the guidelines outlined by Cohen (1988) effect sizes of 0.02, 0.15 and 0.35 point to low,

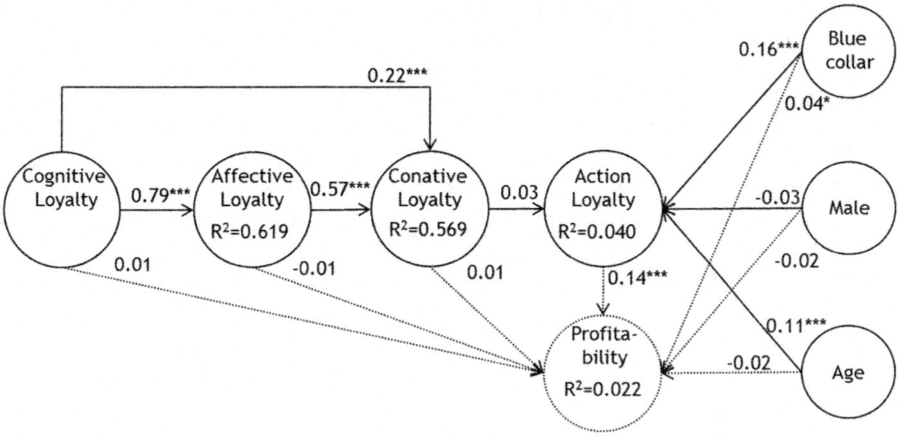

Fig. 3.3 Results of the analyses on the impact on profitability (Step 2)

medium and strong effects of the exogeneous on the endogeneous construct. Effect sizes below 0.02 point to no effect. Hence, we conclude that the effect of action loyalty on customer profitability can, if at all, be deemed as low.

To further test the effects on customer profitability and derive practical implications, we also analyse q^2 effect sizes and performed a blindfolding to assess the predictive relevance of the model (Geisser 1974; Stone 1974). Blindfolding generates the following values for the predictive relevance of constructs: affective loyalty has a Q^2 of 0.422, conative loyalty of 0.456, action loyalty of 0.019 and customer profitability of 0.018. Q^2 is a measure indicating the out-of-sample predictive relevance, i.e., the capability of the model to predict endogenous constructs such as customer profitability. Q^2 values above zero indicate predictive relevance. Hence, we have predictive relevance for all constructs especially for affective and conative loyalty. The q^2 effect sizes can be interpreted correspondingly to the threshold values for the f^2 effect sizes, i.e., values of 0.02 point to a low predictive relevance of a specific exogenous on an endogenous construct. We calculated the q^2 effect size of action loyalty on customer profitability by means of estimating two models, one including the action loyalty construct and one excluding the action loyalty construct. The following formula building on the Q^2 values then generates the q^2 value: $(Q^2_{included} - Q^2_{excluded})/(1 - Q^2_{included})$. The q^2 effect size of action loyalty on customer profitability is 0.016 ((0.018–0.002)/(1–0.018)), again indicating a low, if at all, predictive relevance.

In sum, the model well explains and predicts affective and conative loyalty, yet, performs poor in terms of explaining and predicting action loyalty and customer profitability. Hence, in deriving managerial implications, we are able to derive recommendations to drive affective and conative loyalty. The effect of these measures on action related aspects of loyalty and finally on hard financial data is however rather weak if looked at from a cross-sectional perspective.

3.5 Summary and Conclusions

The four stage model theoretically is highly appealing and some of its key implications are valid in the financial industry. There are strong effects of cognitive on affective and of cognitive and affective on conative loyalty; the model explains well and predicts these constructs in the financial sector. Corresponding to former studies the relationship between conative and action loyalty is, however, the weakest (Evanschitzky and Wunderlich 2006; Harris and Goode 2004; Sivadas and Baker-Prewitt 2000). The former is not fully surprising given the correlation between intentions and action in general is not perfect (Sun and Morwitz 2010). Moreover, the notion of individual loyalty stages which each affect the construct on the next (higher) level or stage is not fully supported in our study in the financial sector.

This corresponds to findings of other authors highlighting that customer loyalty may be more complex than depicted by the linear, sequential structure of the four stage loyalty model (Agustin and Singh 2005; Han et al. 2008). Moreover, there are no significant effects between the cognitive, affective and conative loyalty dimensions on customer profitability. The only construct showing at least a significant path—yet a very low effect—on profitability is action loyalty. Even this construct does not strongly contribute to explaining and predicting profitability. Profitability is influenced by many other variables than customer loyalty, such as competitive and environmental effects (Zeithaml 2000). Hence, the following recommendations derived for managers are to be understood as investments into loyalty dimensions that should have a long-term impact on future success, yet do not trigger immediate positive paybacks in terms of cross-selling and customer profitability on the individual customer level (see also the findings of Rust and Zahorik 1993). At least, investments will trigger conative loyalty and therewith, recommendation of the financial products to friends and relatives, which is a likewise interesting endogenous construct in the financial sector, taking the high acquisition costs of customers into account.

In order to understand what drives recommendation, as well as affective loyalty of customers (i.e., satisfaction and trust), the following cognitive loyalty aspects are of specific relevance in the financial sector. First, the hard facts in the competitive environment count, namely being ahead of the competition and offering the best possible value for money. Second, from the SERVQUAL dimensions, two show a significant impact: reliability and responsiveness (see also Avkiran 1999). Hence, it seems to be about offering individual service, i.e., low waiting times before speaking to consultants and staff that are willing to help and offer reliable consulting. Assurance, empathy and tangibles do not play a significant role in driving up affective and conative loyalty.

Our study in the financial sector finds only partial support for one of the most prominent models in loyalty research, Oliver's (1999) four stage loyalty framework. While we are able to confirm most of the direct relationships in the loyalty cascade, namely, from cognitive to affective loyalty, and affective to conative loyalty, there is no full mediation of constructs between the different stages. Furthermore, our findings do not support the notion that intentions indeed fully reflect actual behavior as has been questioned by other authors in the field (Oliva et al. 1992; Rust et al. 1995). Hence, we call for further research on the cascade model. Moreover, there is only a very small effect from loyalty on profitability, namely, from action loyalty on profitability. This minor effect is rather in line with earlier research which found that action loyalty explains 10% of the variations of customer profitability (Helgesen 2006). Our research questions the assumed link between loyalty and profitability on the individual customer level. Nonetheless, we suggest more research to test this link to profitability which might also look at more operational performance measures such as product market outcomes (sales growth), cost efficiency and productivity outcomes or may take an information cost perspective (Richter et al. 2017; Richter 2014).

For management in the financial sector, the recommendation is to carefully identify the profitable customer segments, as there is no clear cut effect from loyalty to profitability. For these segments, actionable strategies to drive loyalty in different stages are a fruitful approach in developing customer retention plans and increasing the financial outcome of customer relationships. For researchers, PLS-SEM is a useful tool in the domain due to the explorative nature of the study in analyzing competing loyalty cascades and their impact on profitability in a dynamic and complex context.

The above findings need to be considered in light of the following limitations of our empirical study regarding the sample, time frame and data collected. First, as in every empirical research design, the operationalization of our variables involves weaknesses and is not a perfect reflection of business realities. Second, our sampling focused on the customers of one bank only. To further validate findings, research involving a broader set of financial institutions would be desirable. This is especially true as research suggests that the chain of effects in the loyalty cascade may differ depending on the product under study (Olsen 2002). Third, we performed a cross-section study focusing on 1 year only. In order to derive in-depth insights into the development of customer profitability and the development of individual loyalty dimensions over time, a panel design would be desirable collecting information from individual customers over a longer period. Fourth, future studies may extend the empirical analyses by considering advanced PLS-SEM techniques (Hair et al. 2018, in press) such as the importance-performance map analysis (Ringle and Sarstedt 2016), the FIMIX-PLS and PLS-POS methods to uncover unobserved heterogeneity (e.g., Becker et al. 2013; Hair et al. 2016; Matthews et al. 2016), and the PLS multigroup analysis (Chin 2003; Sarstedt et al. 2011) to generate further differentiated findings and conclusions.

Acknowledgements This chapter uses the statistical software SmartPLS 3 (http://www.smartpls.com). Ringle acknowledges a financial interest in SmartPLS.

References

Agustin, C., & Singh, J. (2005). Curvilinear effects of consumer loyalty determinants in relational exchanges. *Journal of Marketing Research, 42*(1), 96–108.
Ahrholdt, D. C., Gudergan, S. P., & Ringle, C. M. (2016). Enhancing service loyalty. *Journal of Travel Research, 56*(4), 436–450.
Anderson, E. W., & Mansi, S. A. (2009). Does customer satisfaction matter to investors? Findings from the bond market. *Journal of Marketing Research, 46*(5), 703–714.
Anderson, E. W., Fornell, C., & Lehmann, D. R. (1994). Customer satisfaction, market share, and profitability: Findings from Sweden. *Journal of Marketing, 58*(3), 53–66.
Anderson, E. W., & Mittal, V. (2000). Strengthening the satisfaction-profit chain. *Journal of Service Research, 3*(2), 107–120.
Arasli, H., Mehtap-Smadi, S., & Turan Katircioglu, S. (2005). Customer service quality in the Greek Cypriot banking industry. *Managing Service Quality, 15*(1), 41–56.
Athanasopoulou, P. (2009). Relationship quality: A critical literature review and research agenda. *European Journal of Marketing, 43*(5/6), 583–610.
Aurier, P., & N'Goala, G. (2010). The differing and mediating roles of trust and relationship commitment in service relationship maintenance and development. *Journal of the Academy of Marketing Science, 38*(3), 303–325.
Avkiran, N. K. (1994). Developing an instrument to measure customer service quality in branch banking. *International Journal of Bank Marketing, 12*(6), 10–18.
Avkiran, N. K. (1999). Quality customer service demands human contact. *International Journal of Bank Marketing, 17*(2), 61–74.
Avkiran, N. K. (2002). Credibility and staff conduct make or break bank customer service quality. *Journal of Asia-Pacific Business, 3*(3), 73–91.
Becker, J.-M., Rai, A., Ringle, C. M., & Völckner, F. (2013). Discovering unobserved heterogeneity in structural equation models to avert validity threats. *MIS Quarterly, 37*(3), 665–694.
Blau, P. (1964). *Exchange and power in social life*. New York: Wiley.
Bolton, R. N. (1998). A dynamic model of the duration of the customer's relationship with a continuous service provider: The role of satisfaction. *Marketing Science, 17*(1), 45–65.
Bolton, R. N., & Lemon, K. N. (1999). A dynamic model of customers' usage of services: Usage as an antecedent and consequence of satisfaction. *Journal of Marketing Research, 36*(2), 171–186.
Borsboom, D., Mellenbergh, G. J., & Van Heerden, J. (2003). The theoretical status of latent variables. *Psychological Review, 110*(2), 203.
Briggs, E., & Grisaffe, D. (2009). Service performance-loyalty intentions link in a business-to-business context: The role of relational exchange outcomes and customer characteristics. *Journal of Service Research, 13*(1), 37–51.
Brunner, T. A., Stöcklin, M., & Opwis, K. (2008). Satisfaction, image and loyalty: New versus experienced customers. *European Journal of Marketing, 42*(9/10), 1095–1105.
Cannon, J. P., & Perreault, W. D., Jr. (1999). Buyer-seller relationships in business markets. *Journal of Marketing Research, 36*(4), 439–460.
Carrillat, F. A., Jaramillo, F., & Mulki, J. P. (2009). Examining the impact of service quality: A meta-analysis of empirical evidence. *Journal of Marketing Theory & Practice, 17*(2), 95–110.
Chaudhuri, A., & Holbrook, M. B. (2001). The chain of effects from brand trust and brand affect to brand performance: The role of brand loyalty. *Journal of Marketing, 65*(April), 81–93.
Chin, W. W. (2003). A permutation procedure for multi-group comparison of PLS models. In M. Vilares, M. Tenenhaus, P. S. Coelho, V. E. Vinzi, & A. Morineau (Eds.), *Focus on customers: Proceedings of the 3rd international symposium on PLS and related methods (PLS'03)* (Vol. 3, pp. 33–43). Paris: Decisia.
Chin, W. W. (2010). How to write up and report PLS analyses. In V. E. Vinzi, W. W. Chin, J. Henseler, & H. Wang (Eds.), *Handbook of partial least squares: Concepts, methods and applications* (Vol. 2, pp. 655–690). Heidelberg: Springer.

Chiou, J.-S., & Droge, C. (2006). Service quality, trust, specific asset investment, and expertise: Direct and indirect effects in a satisfaction-loyalty framework. *Journal of the Academy of Marketing Science, 34*(4), 613–627.

Cohen, J. (1988). *Statistical power analysis for the behavioral sciences* (2nd ed.). Hillsdale: Lawrence Erlbaum Associates.

Coltman, T., Devinney, T. M., Midgley, D. F., & Venaik, S. (2008). Formative versus reflective measurement models: Two applications of formative measurement. *Journal of Business Research, 61*(12), 1250–1262.

Cronin, J. J., Jr., & Taylor, S. A. (1992). Measuring service quality: A reexamination and extension. *Journal of Marketing, 56*(3), 55–68.

Crosby, L. A., Evans, K. A., & Cowles, D. (1990). Relationship quality in services selling: An interpersonal influence perspective. *Journal of Marketing, 54*(3), 68–81.

Dagger, T. S., & O'Brien, T. K. (2010). Does experience matter? Differences in relationship benefits, satisfaction, trust, commitment and loyalty for novice and experienced users. *European Journal of Marketing, 44*(9/10), 1528–1552.

Day, G. S. (1969). A two-dimensional concept of brand loyalty. *Journal of Advertising, 9*(3), 29–35.

Delgado-Ballester, E., & Munuera-Alemán, J. L. (2001). Brand trust in the context of consumer loyalty. *European Journal of Marketing, 35*(11/12), 1238–1258.

Di Benedetto, C. A., & Kim, K. H. (2016). Customer equity and value management of global brands: Bridging theory and practice from financial and marketing perspectives. Introduction to a Journal of Business Research special section. *Journal of Business Research, 69*(9), 3721–3724.

Diamantopoulos, A., & Winklhofer, H. M. (2001). Index construction with formative indicators: An alternative to scale development. *Journal of Marketing Research, 38*(2), 269–277.

Diamantopoulos, A., Sarstedt, M., Fuchs, C., Wilczynski, P., & Kaiser, S. (2012). Guidelines for choosing between multi-item and single-item scales for construct measurement: A predictive validity perspective. *Journal of the Academy of Marketing Science, 40*(3), 434–449.

Dick, A. S., & Basu, K. (1994). Customer loyalty: Toward an integrated conceptual framework. *Journal of the Academy of Marketing Science, 22*(2), 99–113.

Doney, P. M., & Cannon, J. (1997). An examination of the nature of trust in buyer-seller relationships. *Journal of Marketing, 61*(April), 35–51.

Doz, Y. L., & Prahalad, C. K. (1991). Managing DMNCs: A search for a new paradigm. *Strategic Management Journal, 12*, 145–164.

Du, R. Y., Kamakura, W. A., & Mela, C. F. (2007). Size and share of customer wallet. *Journal of Marketing, 71*(2), 94–113.

Dyer, J. H. (1996). Specialized supplier networks as a source of competitive advantag: Evidence from the auto industry. *Strategic Management Journal, 17*(4), 271–291.

Edvardsson, B., Johnson, M. D., Gustafsson, A., & Strandvik, T. (2000). The effects of satisfaction and loyalty on profits and growth: Products versus services. *Total Quality Management, 11*(7), 917–927.

Eiriz, V., & Wilson, D. (2006). Research in relationship marketing: Antecedents, traditions and integration. *European Journal of Marketing, 40*(3/4), 275–291.

Emerson, R. M. (1976). Social exchange theory. *Annual Review of Sociology, 2*, 335–362.

Evanschitzky, H., & Wunderlich, M. (2006). An examination of moderator effects in the four-stage loyalty model. *Journal of Service Research, 8*(4), 330–345.

Fornell, C., Mithas, S., Morgeson, F. V., & Krishnan, M. S. (2006). Customer satisfaction and stock prices: High returns, low risk. *Journal of Marketing, 70*(1), 3–14.

Fornell, C., Mithas, S., & Morgeson, F. V. (2009). The economic and statistical significance of stock returns on customer satisfaction. *Marketing Science, 28*(5), 820–825.

Garbarino, E., & Johnson, M. S. (1999). The differential roles of satisfaction, trust, and commitment in customer relationships. *Journal of Marketing, 63*(April), 70–87.

Gassenheimer, J. B., Houston, F. S., & Davis, J. C. (1998). The role of economic value, social value, and perceptions of fairness in interorganizational relationship retention decisions. *Journal of the Academy of Marketing Science, 26*(4), 322–337.

Geisser, S. (1974). A predictive approach to the random effect model. *Biometrika, 61*(1), 101–107.

Grönroos, C. (1990). *Service management and marketing*. Toronto: Lexington Books.

Gudergan, S. P., Ringle, C. M., Wende, S., & Will, A. (2008). Confirmatory tetrad analysis in PLS path modeling. *Journal of Business Research, 61*(12), 1238–1249.

Hair, J. F., Sarstedt, M., Matthews, L. M., & Ringle, C. M. (2016). Identifying and treating unobserved heterogeneity with FIMIX-PLS: Part I—Method. *European Business Review, 28*(1), 63–76.

Hair, J. F., Hult, G. T. M., Ringle, C. M., & Sarstedt, M. (2017a). *A primer on partial least squares structural equation modeling (PLS-SEM)* (2nd ed.). Thousand Oaks: Sage.

Hair, J. F., Hult, G. T. M., Ringle, C. M., Sarstedt, M., Richter, N. F., & Hauff, S. (2017b). *Partial Least Squares Strukturgleichungsmodellierung (PLS-SEM): Eine anwendungsorientierte Einführung*. München: Vahlen.

Hair, J. F., Hult, G. T. M., Ringle, C. M., Sarstedt, M., & Thiele, K. O. (2017c). Mirror, mirror on the wall: A comparative evaluation of composite-based structural equation modeling methods. *Journal of the Academy of Marketing Science*, 45(5), 616–632.

Hair, J. F., Sarstedt, M., Ringle, C. M., & Gudergan, S. P. (2018, in press). *Advanced issues in partial least squares structural equation modeling (PLS-SEM)*. Thousand Oaks: Sage.

Hallén, L., Johanson, J., & Seyed-Mohamed, N. (1991). Interfirm adaptation in business relationships. *Journal of Marketing, 55*(2), 29–37.

Hallowell, R. (1996). The relationships of customer satisfaction, customer loyalty, and profitability: An empirical study. *International Journal of Service Industry Management, 7*(4), 27–42.

Han, X., Kwortnik, R. J., & Wang, C. (2008). Service loyalty. An integrative model and examination across service contexts. *Journal of Service Research, 11*(1), 22–42.

Hansen, H., Sandvik, K., & Selnes, F. (2003). Direct and indirect effects of commitment to a service employee on the intention to stay. *Journal of Service Research, 5*(4), 356.

Harris, L. C., & Goode, M. M. H. (2004). The four levels of loyalty and the pivotal role of trust: A study of online service dynamics. *Journal of Retailing, 80*(2), 139–158.

Helgesen, Ø. (2006). Are loyal customers profitable? Customer satisfaction, customer (action) loyalty and customer profitability at the individual level. *Journal of Marketing Management, 22*(3–4), 245–266.

Hennig-Thurau, T., Gwinner, K. P., & Gremler, D. D. (2002). Understanding relationship marketing outcomes: An integration of relational benefits and relationship quality. *Journal of Service Research, 4*(3), 230–247.

Jacoby, J., & Chestnut, R. W. (1978). *Brand loyalty: Measurement and management*. New York: Wiley.

Jacoby, J., & Kyner, D. B. (1973). Brand loyalty vs. repeat purchasing behaviour. *Journal of Marketing Research, 10*(February), 1–9.

Jap, S. D. (1999). Pie-expansion efforts: Collaboration processes in buyer-supplier relationships. *Journal of Marketing Research, 36*(4), 461–475.

Jap, S. D. (2001). Perspectives on joint competitive advantages in buyer-supplier relationships. *International Journal of Research in Marketing, 18*(1/2), 19–35.

Joshi, A. W., & Stump, R. L. (1999). The contingent effect of specific asset investments on joint action in manufacturer-supplier relationships: An empirical test of the moderating role of reciprocal asset investments, uncertainty, and trust. *Journal of the Academy of Marketing Science, 27*(3), 291–305.

Kantsperger, R., & Kunz, W. H. (2010). Consumer trust in service companies: A multiple mediating analysis. *Managing Service Quality, 20*(1), 4–25.

Kumar, V. (2016). Introduction: Is customer satisfaction (ir)relevant as a metric? *Journal of Marketing, 80*(5), 108–109.

Lambe, C. J., Spekman, R. E., & Hunt, S. D. (2000). Interimistic relational exchange: Conceptualization and propositional development. *Journal of the Academy of Marketing Science, 28*(2), 212–225.

Lariviere, B. (2008). Linking perceptual and behavioral customer metrics to multiperiod customer profitability: A comprehensive service-profit chain application. *Journal of Service Research, 11*(1), 3–21.

Lawler, E. J. (2001). An affect theory of social exchange. *American Journal of Sociology, 107*(2), 321–352.

Macneil, I. R. (1978). Contracts: Adjustment of long-term economic relations under classical, neoclassical and relational contract law. *Northwestern University Law Review, 72(6)*, 854–902.

Matthews, L. M., Sarstedt, M., Hair, J. F., & Ringle, C. M. (2016). Identifying and treating unobserved heterogeneity with FIMIX-PLS: Part II—A case study. *European Business Review, 28*(2), 208–224.

Mittal, V., & Katrichis, J. M. (2000). Distinctions between new and loyal customers. *Marketing Research, 12(1)*, 26–32.

Morgan, R. M., & Hunt, S. D. (1994). The commitment-trust theory of relationship marketing. *Journal of Marketing, 58*(July), 20–38.

Mukherjee, A., & Nath, P. (2005). An empirical assessment of comparative approaches to service quality measurement. *Journal of Services Marketing, 19*(3), 174–184.

Niraj, R., Gupta, M., & Narasimhan, C. (2001). Customer profitability in a supply chain. *Journal of Marketing, 65*(3), 1–16.

Nitzl, C., Roldan, J. L., & Cepeda, G. (2016). Mediation analysis in partial least squares path modeling: Helping researchers discuss more sophisticated models. *Industrial Management & Data Systems, 116*(9), 1849–1864.

O'Sullivan, D., & McCallig, J. (2009). Does customer satisfaction influence the relationship between earnings and firm value? *Marketing Letters, 20*(4), 337–351.

Oliva, T. A., Oliver, R. L., & MacMillan, I. C. (1992). A catastrophe model for developing service satisfaction strategies. *Journal of Marketing, 56*(3), 83–95.

Oliver, R. L. (1997). *Satisfaction: A behavioral perspective on the consumer*. New York: McGraw-Hill.

Oliver, R. L. (1999). Whence consumer loyalty? *Journal of Marketing, 63*(Special Issue), 33–44.

Olsen, S. O. (2002). Comparative evaluation and the relationship between quality, satisfaction, and repurchase loyalty. *Journal of the Academy of Marketing Science, 30*(3), 240–249.

Palmatier, R. W., Dant, R., Grewal, D., & Evans, K. R. (2006). Factors influencing the effectiveness of relationship marketing: A meta-analysis. *Journal of Marketing, 70*(October), 136–153.

Papassapa, R., & Miller, K. E. (2007). Relationship quality as a predictor of B2B customer loyalty. *Journal of Business Research, 60*(1), 21–31.

Parasuraman, A., Zeithaml, V. A., & Berry, L. L. (1985). A conceptual model of service quality and its implications for future research. *Journal of Marketing, 49(4)*, 41–50.

Parasuraman, A., Zeithaml, V. A., & Berry, L. L. (1988). SERVQUAL: A multiple-item scale for measuring consumer perceptions of service quality. *Journal of Retailing, 64*(1), 12–40.

Perkins-Munn, T., Aksoy, L., Keiningham, T. L., & Estrin, D. (2005). Actual purchase as a proxy for share of wallet. *Journal of Service Research, 7*(3), 245–256.

Pfeffer, J., & Salanzik, G. R. (1978). *The external control of organizations: A resource dependence perspective*. New York: Harper & Row.

Reichheld, F. F., & Sasser, W. E., Jr. (1990). Zero defections: Quality comes to services. *Harvard Business Review, 68*(5), 105–111.

Reinartz, W. J., & Kumar, V. (2000). On the profitability of long-life customers in a noncontractual setting: An empirical investigation and implications for marketing. *Journal of Marketing, 64*(4), 17–35.

Richter, N. F. (2014). Information costs in International Business: Analyzing the effects of economies of scale, cultural diversity and decentralization. *Management International Review,* 54 (2), 171–193.

Richter, N. F., Cepeda, G., Roldán, J. L., & Ringle, C. M. (2016a). European management research using partial least squares structural equation modeling (PLS-SEM). *European Management Journal, 34*(6), 589–597.
Richter, N. F., Sinkovics, R. R., Ringle, C. M., & Schlägel, C. (2016b). A critical look at the use of SEM in international business research. *International Marketing Review, 33*(3), 376–404.
Richter, N. F., Schmidt, R., Ladwig, T. J., Wulhorst, F. (2017). A critical perspective on the measurement of performance in the empirical multinationality and performance literature. *Critical Perspectives on International Business*, 13(2), 94–118.
Ringle, C. M., & Sarstedt, M. (2016). Gain more insight from your PLS-SEM results: The importance-performance map analysis. *Industrial Management & Data Systems, 116*(9), 1865–1886.
Ringle, C. M., Wende, S., & Becker, J.-M. (2015). *SmartPLS 3*. Boenningstedt: SmartPLS.
Rust, R. T., & Zahorik, A. J. (1993). Customer satisfaction, customer retention, and market share. *Journal of Retailing, 69*(2), 193–215.
Rust, R. T., Zahorik, A. J., & Keiningham, T. L. (1995). Return on quality (ROQ): Making service quality financially accountable. *Journal of Marketing, 59*(2), 58–70.
Sarstedt, M., Henseler, J., & Ringle, C. M. (2011). Multigroup analysis in partial least squares (PLS) path modeling: Alternative methods and empirical results. In M. Sarstedt, M. Schwaiger, & C. R. Taylor (Eds.), *Advances in international marketing* (pp. 195–218). Bingley: Emerald.
Sarstedt, M., Diamantopoulos, A., & Salzberger, T. (2016a). Should we use single items? Better not. *Journal of Business Research, 69*(8), 3199–3203.
Sarstedt, M., Hair, J. F., Ringle, C. M., Thiele, K. O., & Gudergan, S. P. (2016b). Estimation issues with PLS and CBSEM: Where the bias lies! *Journal of Business Research, 69*(10), 3998–4010.
Sarstedt, M., Ringle, C. M., & Hair, J. F. (2017). Partial least squares structural equation modeling. In C. Homburg, M. Klarmann, & A. Vomberg (Eds.), *Handbook of market research*. Heidelberg: Springer.
Seppänen, R., Blomqvist, K., & Sundqvist, S. (2007). Measuring inter-organizational trust—A critical review of the empirical research in 1990–2003. *Industrial Marketing Management, 36*(2), 249–265.
Sierra, J. J., & McQuitty, S. (2005). Service providers and customers: Social exchange theory and service loyalty. *Journal of Services Marketing, 19*(6), 392–400.
Singh, J., & Sirdeshmukh, D. (2000). Agency and trust mechanisms in consumer satisfaction and loyalty judgments. *Journal of the Academy of Marketing Science, 28*(1), 150–167.
Sirdeshmukh, D., Singh, J., & Sabol, B. (2002). Consumer trust, value, and loyalty in relational exchanges. *Journal of Marketing, 66*(1), 15–37.
Sivadas, E., & Baker-Prewitt, J. L. (2000). An examination of the relationship between service quality, customer satisfaction, and store loyalty. *International Journal of Retail & Distribution Management, 28*(2), 73–82.
Stone, M. (1974). Cross-validatory choice and assessment of statistical predictions. *Journal of the Royal Statistical Society, Series B, 36*(2), 111–147.
Storbacka, K., Strandvik, T., & Grönroos, C. (1994). Managing customer relationships for profit: The dynamics of relationship quality. *International Journal of Service Industry Management, 5*(5), 21–38.
Sun, B., & Morwitz, V. G. (2010). Stated intentions and purchase behavior: A unified model. *International Journal of Research in Marketing, 27*(4), 356–366.
Tenenhaus, M., Vinzi, V. E., Chatelin, Y.-M., & Lauro, C. (2005). PLS path modeling. *Computational Statistics & Data Analysis, 48*(1), 159–205.
Thibaut, J. W., & Kelley, H. H. (1959). *The social psychology of groups*. New York: Wiley.
Williamson, O. E. (1975). *Markets as hierarchies: Analysis and antitrust implications*. New York: The Free Press.
Williamson, O. E. (1985). *The economic institutions of capitalism*. New York: The Free Press.

Zeithaml, V. A. (2000). Service quality, profitability, and the economic worth of customers: What we know and what we need to learn. *Journal of the Academy of Marketing Science, 28*(1), 67–85.

Zhao, X., Lynch, J. G., & Chen, Q. (2010). Reconsidering Baron and Kenny: Myths and truths about mediation analysis. *Journal of Consumer Research, 37*(2), 197–206.

Anne-Kathrin Hegner-Kakar is a Ph.D. student at the Hamburg University of Technology (TUHH) and Head of Fashion & Lifestyle (Germany) at GfK SE. She has strong expertise in consulting companies from different sectors including (fashion) retail, finance, automotive and tech. Having worked in the market research industry since 2006, she has significant experience across a range of research topics including customer relationship management, shopper behavior, segmentations and product optimization. Her research focus is on customer loyalty.

Nicole Franziska Richter is an Associate Professor of International Business at the University of Southern Denmark. Her research focus is on international and strategic management. She received her Ph.D. (on Internationalization and Firm Performance) from University of Hamburg (in 2009), awarded with the "Wolfgang-Ritter"-Award for excellent research (in 2013). She received her habilitation (on Determinants of Success in International Business) from the Hamburg University of Technology (TUHH) (in 2015). Her list of publications covers more than 30 publications in the field of international and strategic management, mostly in peer reviewed academic journals. She has gathered rich expertise in international marketing research as she has been working as a consultant in the marketing research industry for several years.

Christian M. Ringle is a Chaired Professor of Management at the Hamburg University of Technology (TUHH) and a Conjoint Professor at the Faculty of Business Law at the University of Newcastle. His research addresses the management of organizations, strategic and human resource management, marketing, and quantitative methods for business and market research. His research in these fields has been published in well-known journals such as *Annals of Tourism Research, International Journal of Research in Marketing, Information Systems Research, Journal of Business Research, Journal ofService Research, Journal of the Academy of Marketing Science, Long Range Planning, MIS Quarterly, Organizational Research Methods*, and *Tourism Management*. He is a cofounder of SmartPLS (http://www.smartpls.com), a statistical software tool with a graphical user interface for partial least squares structural equation modeling (PLS-SEM). More information about Christian M. Ringle can be found at http://www.tuhh.de/hrmo/team/prof-dr-c-m-ringle.html.

Chapter 4
Corporate Reputation: The Importance of Service Quality and Relationship Investment

Lăcrămioara Radomir and Alan Wilson

Abstract Having a strong reputation is a desirable resource for banks which must face today's competitive business environment. This chapter aims to (1) investigate the contribution of service quality and relationship investment to building a good reputation and (2) show that the efforts made by banks in maintaining customer relationships ultimately translate into positive outcomes, such as relationship loyalty. Data collected from banking services consumers from two countries were analysed using partial least squares (PLS). Based on 510 responses from Romanian consumers and 525 responses from UK consumers, the proposed framework is tested separately and causal relationships are then compared to examine whether path estimates are invariant across the two countries. Finally, the data is pooled to increase the generalizability of results. Perceptions of both service quality and relationship investment have a significant contribution to building sympathy towards banks and in developing favourable perceptions about their competence. Furthermore, perceived relationship investment acts as a mediator in the relationship between overall service quality and corporate reputation. Finally, this study reinforces the importance of perceived relationship investment and corporate reputation in developing long-lasting relationships by demonstrating the willingness of customers to reciprocate. The findings extend previous knowledge in the field by emphasising the importance of perceived relationship investment in building customers' sympathy towards service companies as well as favourable perceptions regarding their competence. In addition, the framework herein proposed acknowledges the mediating role of perceived relationship investment between service quality perceptions and corporate reputation. Furthermore, the study builds on signalling

---

L. Radomir (✉)
Department of Marketing, Faculty of Economics and Business Administration, Babeş-Bolyai University, Cluj-Napoca, Romania
e-mail: lacramioara.radomir@econ.ubbcluj.ro

A. Wilson
Department of Marketing, Strathclyde Business School, University of Strathclyde, Glagow, UK
e-mail: alan.wilson@strath.ac.uk

N.K. Avkiran, C.M. Ringle (eds.), *Partial Least Squares Structural Equation Modeling*, International Series in Operations Research & Management Science 267, https://doi.org/10.1007/978-3-319-71691-6_4

and social exchange theories to show that banks' endeavour to increase their customers' quality perceptions enhance the benefits of banking institutions. Altogether, results reported in this study are of high relevance from both a theoretical and managerial perspective. Future studies may test the proposed framework in other cultures and explore the mediating role of perceived relationship investment between other factors and corporate reputation. Despite the interest in the two concepts of corporate reputation and perceived relationship investment in either the relationship marketing literature or the corporate reputation literature, the relationship between the two constructs has been only theoretically suggested. Against this background, this chapter is the first to investigate empirically the impact of perceived relationship investment on corporate reputation. It is also the first to assess the potential role of perceived relationship investment in mediating the effect of service quality on corporate reputation.

Keywords Corporate reputation · Perceived relationship investment · Signalling theory · Social exchange theory · Partial least square path modelling · PLS-SEM

4.1 Introduction

The objective of this chapter is to demonstrate that variables specific to signalling and social exchange theories can jointly contribute to the long-term success of banking institutions. To accomplish this purpose, the present chapter simultaneously considers the contribution that perceptions of both service quality and relationship investment have in explaining end-use customers' perceptions of banks' reputation. These, in turn, are further considered essential in enhancing customers' willingness to reciprocate by demonstrating increased relationship loyalty towards their banks.

A set of three reasons give ground for the present study. First, from a managerial perspective, research on corporate reputation has become more important than ever (Shamma 2012). The economic environment in which companies operate is increasingly competitive. Consequently, product and service prices, distribution channels, and offer diversity can hardly guarantee the competitive advantage that can help companies in their endeavour to maintain their positions in the market. Companies must therefore invest in intangible resources that are difficult to imitate by competitors. Of these, corporate reputation was the focus of both researchers and business practitioners during the last decade (Alniacik et al. 2012; Shamma 2012) for the following two reasons.

A favourable corporate reputation is acknowledged as important for its influence on the behaviour of various stakeholders (Chun 2005), which ultimately translates into relevant outcomes for companies. For instance, corporate reputation is an important factor in the process of attracting talented and high-quality employees (Alniacik et al. 2012; Vidaver-Cohen 2007; Walsh and Wiedmann 2004) and retaining them (Chun 2005; Feldman et al. 2014). Similarly, other researchers emphasise the role of corporate reputation in attracting potential investors (Feldman et al. 2014; Gardberg 2006; Walsh and Wiedmann 2004). A good reputation of a given company facilitates an increase in the number of its customers (Walsh and Wiedemann 2004) by influencing customers' decisions regarding a set of products and/or services (Alniacik et al. 2012; Vidaver-Cohen 2007). Corporate reputation may lead to customers' decision to increase relationship depth and breadth and easily accept new products and/or services (Alniacik et al. 2012; Jeng 2011; Zhang 2009). In addition, it contributes to building better relationships between buyers and sellers by increasing customers' satisfaction (Chun 2005; Walsh and Beatty 2007) and trust (Gardberg 2006; Walsh and Beatty 2007). Finally, a strong corporate reputation influences customers' purchase intentions (Alniacik et al. 2012), their loyalty towards the company (Chun 2005; Schwaiger 2004; Walsh and Beatty 2007), and their decisions to engage in positive word of mouth communication (Feldman et al. 2014; Walsh and Beatty 2007).

Moreover, considering its influence on the behaviour of stakeholders of a company, a favourable corporate reputation can help companies differentiate themselves from their competitors, achieve corporate success (Alniacik et al. 2012), and maintain a sustainable competitive advantage (Eberl 2010; Feldman et al. 2014; Shamma 2012; Schwaiger 2004; Vidaver-Cohen 2007). In light of the recent financial crises, corporate collapses, and scandals, corporate reputation has become even more important as it represents a source of goodwill (Feldman et al. 2014) and can successfully safeguard companies (Shamma 2012). Given all the benefits mentioned above, corporate reputation is regarded as one of the most valuable intangible assets (Schwaiger 2004; Shamma 2012; Shamma and Hassan 2009; Vidaver-Cohen 2007).

Of all stakeholder groups, end-use customers are considered the most important stakeholders (Shamma and Hassan 2009; Terblanche 2014; Walsh et al. 2009a, b) for which the companies work. A company's revenues, and thus its financial performance, are largely dependent on end-use customers (Shamma and Hassan 2009, Walsh et al. 2009a), who may either decide to opt or not to opt for its products. Moreover, it is argued that customers can successfully influence other customers' (Walsh et al. 2009a) as well as other stakeholders' perceptions regarding a company (Shamma and Hassan 2009). Customers are today more demanding, have greater expectations than in the past and look for high-quality customer service and, innovative products and/or services (Shamma and Hassan 2009). Therefore, although often neglected in past studies (Terblanche 2014; Walsh et al. 2009a, b), end-use customers should be treated as a separate and particularly important stakeholder group (Walsh et al. 2009a), which influences the reputation and long-term survival of companies (Shamma and Hassan 2009).

Second, the research on the relationship between service quality and corporate reputation has mainly followed two distinct approaches. On the one hand, drawing on signalling theory, it is suggested that the reputation of companies is a signal of their offering quality. As a result, product and/or service quality is considered either a dimension of corporate reputation or its consequence. On the other hand, product and/or service quality has been argued to be a driver of reputation and not the other way round. Though less embraced in the literature, this perspective holds that once customers experience the products and/or services of a company, they can form quality perceptions. In this respect, Jing and Lei (2007) conclude that reputation exerts an influence on perceived quality only before a purchase and that customers' quality perceptions act as drivers of reputation after the purchase. Similarly, Wang et al. (2003) regard enhanced reputation as one of the benefits assured by a high-quality offering. Jeng (2011) also supports this approach and argues that reputation can only influence customers' expectations, i.e. 'customers' perceptions of the quality of new services' (p. 854). Despite extensive research on corporate reputation, scholars have rather neglected the view that quality can act as a signal of reputation when customers experience a service company's offering.

Third, the importance of perceived relationship investment is well acknowledged in relationship marketing literature and social exchange theory. In essence, it is argued that ongoing relationships are built on the norm of reciprocity (Palmatier et al. 2006; Smith and Barclay 1997; De Wulf et al. 2001). Accordingly, if companies are perceived to make efforts to maintain relationships with their customers, then customers are willing to reciprocate in order for the relationship to last. Researchers have therefore investigated the impact that quality may have on perceived relationship investment and on variables reflecting customers' loyalty towards product and/or service providers (e.g. De Wulf et al. 2003; Liang and Wang 2006). Nevertheless, there is little evidence on the influence of perceived relationship investment on companies' reputation. Specifically, this relationship has been only theoretically suggested in the study of Jarvenpaa et al. (1999), thus pointing towards a gap in the corporate reputation literature.

Considering the aforementioned issues, several contributions to the body of knowledge ensue in this study. First, this study aims to complement existing literature by acknowledging the value of end-use customers in the long-term success of banking institutions. Second, this study argues in favour of the perspective that service quality is a signal of corporate reputation. Subsequently, customers' sympathy towards their financial providers and their favourable perceptions of banks' competence are considered as consequences of banks' investments in the quality of their offerings. Third, this study aims to confirm previous findings and argues that service quality signals the investments made by banks to maintain the relationships with their customers. Extending prior research, inferences based on the social exchange theory are made to substantiate that banks' endeavours to maintain the relationships with their customers are of high importance in building favourable reputation perceptions. Finally, this study aims to confirm that customers' behaviour, as expressed by their loyalty to the relationship with banking institutions, is in accordance with their beliefs and attitudes.

This study collects data in two countries to test the proposed framework. It is argued that such an approach is necessary to increase the generalizability of findings and prove the applicability of a framework across different cultural contexts (Mackenzie 2003; Rosenzweig 1994; Steenkamp and Baumgartner 1998). The following two arguments determined the selection of the countries, namely Romania and the UK. First, despite both countries being located in Europe, the length of existence of a banking system complying with the market economy principles differs between the two countries. Consequently, competition among banking institutions as well as end-use customers' experience with financial services in the two countries are also different between the two countries. Second, Romania and UK represent different cultures and differ in terms of several formal institutions. By departing from the findings of Gardberg (2006), it is important to show that the antecedents and consequences of corporate reputation considered in the framework proposed in this study, are valid across both cultures. Since differences between countries are acknowledged to influence the relationships between the focal construct and its antecedents and/or consequences (Schlägel and Sarstedt 2016), the decision to test the proposed framework with data from the two countries is considered appropriate. A similar approach in the corporate reputation literature was also adopted by Falkenreck and Wagner (2010), Jin et al. (2008), Ruiz et al. (2016), among others.

The remainder of the chapter continues with a discussion on the theoretical background of the research framework proposed herein and graphically synthesised in Fig. 4.1. Section 4.2.5.1 describes the methodology adopted to collect and analyse the data. Following this, findings from the research conducted among the customers of the Romanian and the UK banks are reported. The discussion section emphasises both theoretical and managerial implications and the avenues for future research.

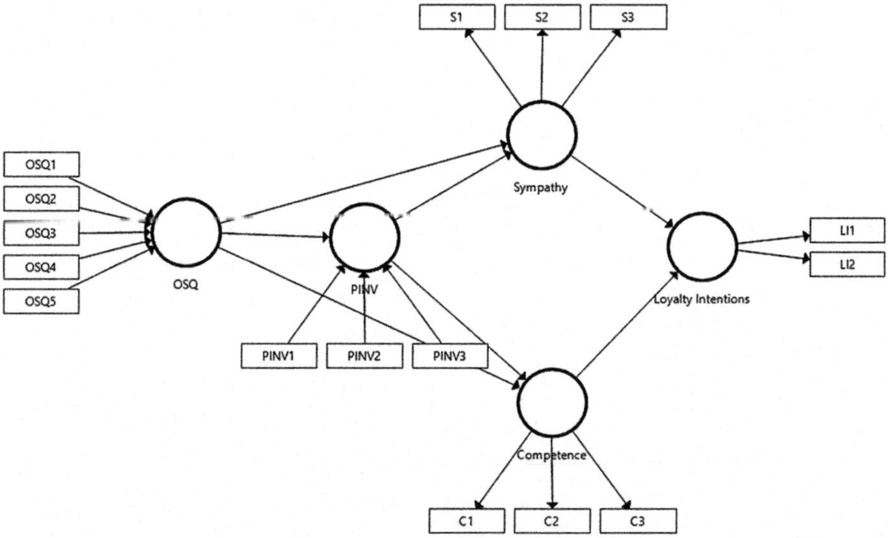

Fig. 4.1 Conceptual model and research hypotheses. Notes: OSQ—perceived overall service quality; PINV—perceived relationship investment; OSQ, PINV, Sympathy, Competence and Loyalty intentions are unobservable first order constructs; OSQ1 … OSQ5 are observable composite indicators corresponding to the OSQ composite construct; PINV1 … PINV3 are observable composite indicators corresponding to the PINV composite construct; S1 … S3 are observable reflective indicators corresponding to the Sympathy reflective construct; C1 … C3 are observable reflective indicators corresponding to the Competence reflective construct; L1, L2 are observable reflective indicators corresponding to the Loyalty intentions reflective construct

4.2 Theoretical Reasoning and Research Framework

4.2.1 Corporate Reputation

Although there are different views on the conceptualisation of reputation, there is consent on its perceptual nature (Vidaver-Cohen 2007). Broadly, reputation can be defined from a multiple stakeholder perspective. Specifically, reputation is defined as an overall perception or evaluation of an organisation made by its constituents over time (Feldman et al. 2014; Shamma and Hassan 2009; Vidaver-Cohen 2007). Despite being appealing, this approach has been recently criticised by Walker (2010) and Walker and Dyck (2014), who doubt the adequacy of an overall reputation rating reflecting the evaluations of all stakeholders on every possible criterion. Therefore, researchers are advised to examine and consequently define reputation from the perspective of a single stakeholder group.

In the present study, following Walsh and Beatty (2007) and Schwaiger (2004), reputation is conceptualised as an attitude construct reflecting customers' judgements of a service company based on their past direct experiences and on the information received about the company. Attitude, in line with Schwaiger's (2004) approach, 'denotes subjective, emotional, and cognitive based mindsets' (p. 49). Accordingly, reputation is considered a two-dimensional construct, which consists of an affective and a cognitive component referred to as sympathy, and competence, respectively.

4.2.2 Signalling Theory

Signalling theory is a popular approach among consumer researchers (Boulding and Kirmani 1993), animal behaviourists, human behavioural ecologists, biologists, psychologists, and economists (Dunham 2011). In the context of the present study, the prominent position of signalling theory was acknowledged in both management (Connelly et al. 2011) and marketing research (Dunham 2011) and was identified among the most common theories in the corporate reputation literature by Walker (2010). Signalling theory provides a basis for describing and understanding behaviour when entities involved in a market interaction have access to or possess different information (Connelly et al. 2011). Information asymmetry may occur between companies and their employees, between company managers and potential investors in the company, as well as between a company and consumers (Stiglitz 2002). Nelson (1970) asserts that information asymmetry poses problems for experience products since sellers know the quality of their offer prior to the transaction, while customers can only assess it after product purchase and usage. Hence, signalling is deemed effective in product markets with quality sensitive consumers (Kirmani and Rao 2000), who lack important information about companies' offerings, so they can easily distinguish between high- and low-quality product sellers (Boulding and

Kirmani 1993) and make better decisions (Connelly et al. 2011). In such settings, companies want to decrease the information asymmetry between them and the consumers. Therefore, companies assume the role of the signaller who communicates the unobservable quality by sending observable signals (Kirmani and Rao 2000) to emphasise the positive attributes of the organisation and its offerings (Connelly et al. 2011). Such actions undertaken by companies are expected to alter consumers' behaviour (Stiglitz 2002).

Banking services possess experience attributes, which can only be assessed after the customers experience the service. Consequently, it is assumed that banking institutions hold more and different types of information about their service offerings when compared to the consumers. Building on this reasoning, most researchers in the corporate reputation literature maintain that financial service providers need to provide observable signals to the consumers in order to communicate the imperceptible quality of their offerings. Among these signals, corporate reputation is deemed a valuable signal that may be used to provide information about the quality of service offering (Walsh et al. 2014). From this perspective, building a good reputation is important, given consumers' uncertainty about the quality of banking services. Nevertheless, the relationship between service quality and reputation is far from being that straightforward. For instance, Podonly and Phillips (1996) suggest that reputation is determined by the value of a company's previous efforts, which in turn is influenced by the quality of the company's offerings. Accordingly, this perspective holds that quality may be used by banks as a signal for their reputation. In fact, as will be argued later in Sect. 4.2.3, consumers develop quality perceptions when they experience a banking service, which may further affect their perceptions about the reputation of their service providers.

4.2.3 Social Exchange Theory

According to the social exchange theory, an individual's decision to put effort into an activity is motivated by expected benefits, i.e. returns (Blau 1964), which are not necessarily immediately observable (Shiau and Luo 2012), but which are thought to justify the costs associated with the investment in the relationship (Liao 2008). Hence, interactions between individuals are considered exchanges that require both inputs in the form of costs and outputs in the form of benefits. Put differently, social exchanges require investment in the relationship for facilitating a sustainable exchange that leads to the expected benefits (Shore et al. 2006). This means that a cyclical reciprocation is expected (Palmatier et al. 2006; Palmatier 2008) where an action undertaken by one party should lead to a response by the other party (Cropanzano and Mitchell 2005). Reciprocity, therefore, is a key concept in the social exchange theory.

Despite its origins in interpersonal relationships and social psychology, social exchange theory is appropriate for evaluating the relationships between buyers and sellers. Consequently, social exchange theory has been extensively referred to in a business-to-business (B2B) context and can be applied to consumer studies as well.

In this study, social exchange implies that actions undertaken by banking institutions to respond to customers' needs lead to favourable customer perceptions. In turn, customers are expected to feel a commitment to respond with a positive signal and engage in behaviours that would improve the performance of their financial service provider. When the efforts of service providers are observable, their actions are perceived as investments in the relationship. This may translate customers' response into positive corporate reputation perceptions. Further, social exchange theory suggests that customers may want to reciprocate and demonstrate that they value the relationship built by showing a predisposition towards maintaining the relationship.

Based on signalling and social exchange theories, this study proposes that perceptions about both service quality and relationship investments of banking institutions are important determinants of consumers' corporate reputation perceptions. The more favourable these perceptions, the more inclined consumers are to respond with a favourable behaviour, as expressed through relationship loyalty. In the present study, the resulting constructs of corporate reputation and relationship loyalty reflect customers' reciprocation in relation to the relationship investments made by their banks as they embody the extent to which customers develop positive beliefs and adopt a favourable behaviour towards their financial service providers.

4.2.4 Antecedents of Corporate Reputation

4.2.4.1 Service Quality

Many researchers interested in the reputation of service companies have referred to the signalling theory to argue their decision to treat corporate reputation as a signal of the quality of services offered to customers. Accordingly, service quality was treated either as a consequence (e.g. Jeng 2008; Loureiro and Kastenholz 2011) or as one of the distinct dimensions of corporate reputation (e.g. Bartikowski et al. 2011; Rindova et al. 2005). Given the asymmetry in the information between exchange partners, this approach posits that it is difficult for consumers to make judgements about experience products that are hard or impossible to evaluate before consumption (Weigelt and Camerer 1988). That is, customers do not have the necessary information about the characteristics of the service company (Connelly et al. 2011). The service companies make efforts to improve their reputation for reducing this asymmetry in information, which in turn improves consumers' perceptions of the quality of services (Feldman et al. 2014). Consequently, reputation is deemed a good predictor of the quality of service companies' offering (Dawar and Parker 1994). In line with this thinking, quality is associated with the unobservable ability of the signaller to satisfy the needs of an outsider who notices the signal (Connelly et al. 2011) and 'reputation can be thought of as an informative sign about the organization's likely behaviour and quality performance' (Feldman et al. 2014, p. 55). Researchers investigating the reputation of banking institutions have undertaken a similar approach. For instance, Bartikowski and Walsh (2011) and Walsh et al.

(2014) conclude that a good reputation serves as a quality signal as well as a signal of companies' future actions. This, in turn, leads consumers to show a favourable behaviour towards their service providers.

Despite this widely accepted perspective, it is herein posited that once consumers gain experience with a service company, they form perceptions regarding the quality of the offering, which in turn, influence the perceived reputation of the service provider. Hence, quality becomes a signal of reputation. This thinking is in line with the suggestions by Keh and Xie (2009) and Jing and Lei (2007) who maintain that reputation acts as a signal of quality only before product purchase and consumption. Similarly, reputation may serve as a quality signal when a service is relatively new to the market and customers may be relatively uninformed (Kirmani and Rao 2000). However, immediately after consumers use a product and experience its real quality, reputation ceases to be a driver and becomes a consequence of quality perceptions. A similar approach was followed by Wang et al. (2003), Schwaiger (2004), Eberl (2010), Sarstedt and Schloderer (2010), among others. For instance, Wang et al. (2003) conclude that both overall service and product quality exert a positive influence on bank reputation. Service quality has also been proven to be a significant driver of both likeability and competence in various industries, such as mobile communications (Eberl 2010), and financial and insurance industries (Schwaiger 2004). The authors referred to earlier do not regard quality as an inherent part of reputation or as its consequence, but as one of the factors that may contribute towards building a reputation for companies.

Overall service quality is defined as customers' perceptions that a company offers innovative and, high-quality services, which are reliable and possess value for the money (Walsh and Beatty 2007; Walsh et al. 2009a). Drawing on propositions from signalling theory, it is herein propounded that customers who already have experience with a particular banking institution rely on current information related to actions of the financial service provider to make overall quality judgements. These, in turn, are expected to influence customers' perceptions regarding the reputation of the bank. For instance, when consumers have experience with a particular bank, the actions undertaken by the institution become easily observable and convey important information to customers. Consequently, customers develop perceptions regarding the reliability of the bank, the quality of services, and the extent to which services may be deemed innovative. Such perceptions, taken together, contribute to customers' overall quality perceptions. Overall quality is assumed to act as a signal of bank reputation, and customers' overall quality perceptions are therefore expected to influence their perceptions concerning both dimensions of corporate reputation. Accordingly, it is hypothesised that:

H1a: The higher customers' overall service quality perceptions, the higher their sympathy towards their financial service provider;

H1b: The higher customers' overall service quality perceptions, the higher their perceptions regarding the competence of their financial service provider.

4.2.4.2 The Mediating Role of Perceived Relationship Investment in the Relationship Between Service Quality and Corporate Reputation

In line with De Wulf et al. (2001), perceived relationship investment is defined as the extent to which customers perceive that their banking service provider allocates resources and makes efforts to maintain and enhance the relationships with regular customers (De Wulf et al. 2001). Companies seeking for reciprocal actions from their customers need to send out signals that emphasise companies' unobservable intentions. Drawing on signalling theory, it is herein argued that service quality plays an important signalling role in the relationship between service providers and their customers. When banking institutions undertake actions intended to offer high-quality services, these are easily observed by their customers who acknowledge their financial service providers' efforts as proofs for the concern about the relationship. Contrarily, when services are perceived of poor quality, perceived relationship investment diminishes. This is because customers may perceive that their financial service providers do not devote the necessary attention, time, effort and resources to maintain the relationship. This thinking is in line with the suggestion of De Wulf et al. (2001) to consider service quality as an antecedent of relationship investment. Later, De Wulf et al. (2003) and Liang and Wang (2006) empirically show that service quality perceptions exert a positive effect on perceived relationship investment. This finding demonstrates that service quality is considered a signal of providers' unobservable intention to maintain the relationship with their customers.

The research framework proposed in this chapter further draws on the social exchange theory, which emphasises the concept of reciprocity in social interactions (Chibucos et al. 2005). In business-to-consumer relationships, reciprocity may suggest that customers' decision to continue their relationship with a service company is determined by their perceptions about the interest of the service provider to develop long-term relationships. Palmatier et al. (2006) argue that companies first need to make an investment of time, effort, and resources to build strong and long-lasting relationships with their customers. Within the social exchange theory, relationship investment is acknowledged as a key factor for ongoing exchange relationships (Smith and Barclay 1997) and setting expectations of reciprocation (De Wulf et al. 2001; Smith and Barclay 1997). After perceiving their service providers' intention to maintain the relationship, customers are expected to develop positive impressions and feelings about the company and to become receptive to relationship maintenance (Bendapudi and Berry 1997).

Perceived relationship investment has not received much attention in the corporate reputation literature. It is only Jarvenpaa et al. (1999) who stated, 'the marketing literature argues that reputation is a valuable asset that requires a long-term investment of resources, effort, and attention to customer relationships' (p. 48). Nevertheless, no empirical support for this relationship has been found. Based on the social exchange theory and the norm of reciprocity, it is here argued that service providers who put efforts to maintain a relationship with their customers receive a positive feedback. In the context of this chapter, customers' positive perceptions of the reputation of their banks are considered a reciprocal consequence of perceived relationship investment.

In line with the above stated, it is proposed that service quality perceptions not only exert a direct impact on corporate reputation perceptions but also an indirect one, through perceived relationship investment. Specifically, perceived relationship investment is expected to partially mediate the relationship between service quality and corporate reputation perceptions. Consequently, the following hypotheses are put forward:

H2: The higher customers' overall service quality perceptions, the higher their perceived relationship investments made by the financial service providers;

H3a: The higher customers' perceptions regarding financial service providers' investments in the relationship, the higher their sympathy towards the financial service providers;

H3b: The higher customers' perceptions regarding financial service providers' investments in the relationship, the higher their perceptions about the competence of the financial service providers;

H4a: Perceived relationship investment positively mediates the relationship between overall service quality perceptions and the affective component of reputation (i.e. sympathy);

H4b: Perceived relationship investment positively mediates the relationship between overall service quality perceptions and the cognitive component of reputation (i.e. competence).

4.2.5 Consequences of Corporate Reputation

4.2.5.1 Relationship Loyalty

Relationship loyalty is conceptualised as customers' intention to continue the relationship with their service providers. To investigate the impact that reputation has on relationship loyalty, the proposed framework builds upon the social exchange theory. Companies do something for their customers and expect them to exhibit a positive behaviour in return (Palmatier et al. 2006). Customers' decision to switch to competitors might lead them to lose the benefits associated with relationship-specific investments made by their service providers (Bendapudi and Berry 1997). Consequently, customers are likely to remain in the relationship. Further, Shiau and Luo (2012) note that perceived reputation should drive intention. As services can be hardly distinguished from one company to another, companies' reputation is expected to act as a differentiator (Walsh and Beatty 2007). Therefore, favourable impressions about the reputation of companies, as induced by investments made by service providers to maintain the relationship with their customers, embody consumers' reciprocation and are assumed positively associated with customers' loyalty.

Previous studies have provided empirical evidence regarding the impact of corporate reputation on customer loyalty. Bartikowski and Walsh (2011), Bontis et al. (2007), and Walsh et al. (2014) showed that customers' positive reputation perceptions drive their loyalty towards banks. More so, in the mobile communication industry, Eberl (2010) empirically established that likeability drives customers' loyalty towards the company. In line with previous findings, it is hypothesised that:

H5a: The higher customers' sympathy towards their financial service providers, the higher their loyalty towards the relationship;

H5b: The higher customers' perceptions about the competence of their financial service providers, the higher their loyalty towards the relationship.

Departing from the above stated, the present study extends previous research by building on the principles of both signalling and social exchange theories. The conceptual framework in Fig. 4.1. shows that perceived overall service quality and perceived relationship investment exert a positive influence on corporate reputation. Moreover, the framework acknowledges the mediating role of perceived relationship investment in the relationship between perceived overall service quality and the two dimensions of corporate reputation. Finally, both sympathy and competence are expected to have a positive effect on customers' intention to continue the relationship with their service provider.

4.3 Method

4.3.1 Study Context and Sample

This study uses data collected from bank customers in Romania and the UK during a one-month period in 2015. Walker (2010) argues that reputation is built in the specific institutional context of companies. Therefore, when researchers aim to generalize their findings, it is important to test the framework in different contexts. As already noted in the Introduction section, the two countries under consideration differ in terms of their banking environment. While competition in the UK banking sector became stronger during the 1970s (Braggion and Ongena 2013), when the deregulation of financial markets began (Davies et al. 2010), it was only at the end of 1990 that the banking reform started in Romania (Doltu 2000). Despite the new two-tier structure of the Romanian banking system (The Central Bank and the commercial banks), the commercial banking sector continued to be mainly state-owned until 1998 (Barisitz 2004). Hence, the banking sector in Romania, which is an emerging market, is still less developed than those in the Western economies, among which is the banking sector in the UK. Therefore, there are differences in the financial behaviour of consumers in the two countries as well.

International Monetary Fund (2013) notifies that, in 2013, the number of commercial bank branches for every 100,000 adults reached 25.19 in the UK, and the number was slightly higher, 31.65, in Romania during the same period. Contrarily, the number of automated teller machines per 100,000 adults in the UK was slightly above the double of that in Romania, i.e. 128.78 as compared to 63.92. Closely related, differences were also reported concerning the number of payment cards issued by banks in the two countries. In the UK, the number of payment cards issued amounted to 159 million in 2014 (The UK Cards Association 2015), while in the Romanian market the number of payment cards issued totalled to approximately 14.45 million (EZB n.d.). The total amount of loans (deposits) with commercial banks by all households represents 66.47% (61.78%) of the Gross Domestic Product in the UK, while the same indicators only reach 16.10% (20.26%) in Romania (International Monetary Fund 2013). As for the ratio of non-performing loans to total gross loans, The World Bank (2013) reports reports it to be 21.87% in Romania and only 3.11% in the UK. Apart from consumers' debt and savings behaviour, notable differences are also found concerning the online banking penetration. The share of individuals using Internet banking in 2014 reached 4% in Romania (Eurostat n.d.), a percentage significantly lower than 53%, which was reported for the UK (Office for National Statistics (UK) n.d.).

To further argue that Romania and the UK have distinct political, legal, or cultural characteristics, the approach of Schlägel and Sarstedt (2016) is adopted and reference to both the formal (rules, laws, and regulations) and informal institutional environment (cultural norms and values) is made for each of the two countries. In terms of formal institutions, it is only one of the six indexes selected, namely the freedom rating (Freedom House 2015a) that points to a similarity between Romania and the UK, both countries being classified as free with respect to the political rights and civil liberties. Contrarily, regarding the independence of media, Romania is only partly free as compared to the UK, which is classified as free by Freedom House (2015b). In terms of the democracy index, Romania ranks among the countries with a flawed democracy (rank 59 out of 167), while the UK, with a rank of 16, has a full democracy regime (The Economist Intelligence Unit 2015). Concerning the adherence to the rule of law, the UK ranked 12 out of 102 countries in 2015, while Romania only reached the 32nd position (World Justice Project 2015). As for the general public perception of corruption in the public sector, Romania is perceived as rather corrupt with a score of 46 on a scale ranging from 0 to 100, while the UK, with a score of 81 according to the corruption perception index, is perceived to have a clean public sector (Transparency International 2015). Finally, Romania and the UK also differ in terms of the perceived political stability and absence of violence/terrorism, as evidenced by the index reported by The World Bank (2015) for the two countries (Romania's score 2015 = 0.20; the UK score 2015 = 0.56).

The informal institutional environment is also different in the two countries, as evidenced by the scores attributed by Hofstede et al. (2010) to five out of the six cultural value dimensions. The UK scores either medium to high or high for masculinity, individualism, and indulgence and exhibits low scores for power distance and uncertainty avoidance. However, Romania is ranked among the most feminine, collectivist, and restraint-scoring societies. Further, compared to the UK, Romania scores very high on power distance and uncertainty avoidance. It is only the long-term orientation index that suggests a similarity between the two countries.

Consumers in the two countries were approached over the Internet and asked to complete an online survey concerning their main financial service provider. The online questionnaire was accessible through a link sent by one of the leading digital market research companies to its panel of consumers in each of the two countries. A total of 1146 completed questionnaires were returned. Responses were screened to identify straightliners, which refers to individuals who had made the same evaluation for 90% or more statements in each series of questions for at least three out of the five blocks of questions in the questionnaire. After the data purification process, a final sample of 1035 usable cases was reached. Of these, 49.5% were Romanian and 50.7% were consumers from the UK. Overall, 50.4% of the respondents were female and 49.6% were male. The average age was 41.81, with most respondents being aged between 30 and 54 years.

4.3.2 Conceptualisation and Operationalisation of Constructs

Each of the five constructs in the proposed model was previously validated in either the relationship marketing or the corporate reputation literature. Following the approach of Schwaiger (2004), corporate reputation was conceptualised as an attitude construct with affective and cognitive components, wherein the former is referred to as sympathy and the later as competence. Perceived overall service quality was operationalised with five observable variables adopted from Walsh and Beatty (2007). Three indicators adopted from De Wulf et al. (2001) measured perceived relationship investment. Customers' intentions to continue the relationship with their banking institutions were assessed with two items adapted from Colgate and Smith (2005) and Dimitriadis (2010).

Sympathy, competence, and loyalty intentions are unobserved reflective variables, while perceived overall service quality and perceived relationship investment have composite indicators. Hence, contrary to Walsh and Beatty (2007) and in line with Schwaiger (2004), perceived overall service quality is considered a composite construct and an antecedent of corporate reputation rather than one of its distinct dimensions. Similarly, perceived relationship investment is not operationalised by reflective indicators as in the study of De Wulf et al. (2001), but as a latent phenomenon defined by its corresponding three indicators. Each of the eighteen observable variables was measured on a seven-point Likert scale, with 1 indicating 'total disagreement' and 7 'total agreement'.

4.3.3 Data Analysis

To test the conceptual model presented in Fig. 4.1 this study employs a variance-based structural equation modelling (SEM) method, namely partial least squares (PLS). The choice for this method can be attributed to the exploratory nature of the framework proposed in this study. As previously noted, the relationship between perceived relationship investment and corporate reputation has not been empirically investigated to date. Furthermore, perceived overall service quality and perceived relationship investment are operationalised as composite constructs. Besides, this study focuses on predicting competence and sympathy by means of overall service quality perceptions and perceived relationship investment. In such circumstances, the PLS method is preferred to covariance-based techniques (Chin 2010; Hair et al. 2011, 2012, 2014; Vinzi et al. 2010).

The SmartPLS 3.2.4 release (Ringle et al. 2015) statistical software is used for analysis. In line with the logical sequence of analyses suggested by Henseler et al. (2009), a two-phase process is followed to evaluate the model. First, the outer model is assessed to establish whether the quality criteria for both reflective and composite measures are met, and to investigate the relationships between the observable and unobservable variables. In the second stage, the inner model is evaluated with the purpose of investigating the relationships between the five unobservable constructs. This sequence in the model assessment process is followed both for the overall sample and the two subsamples that distinguish between the UK and Romanian banking services users.

Following the recommendations of Hair et al. (2014, 2017), the path model is estimated with the path weighting scheme for the inner weights estimation. The significance of composite indicators' weights and path coefficients is tested by applying the bootstrapping procedure with 5000 bootstrap samples; two-sided significant test, with the no sign change option to deal with sign changes; 1035 cases for the overall sample, 510 cases for the Romanian sample and 525 cases for the UK sample. In order to establish whether path coefficients between the five constructs in the conceptual model are invariant across the two samples, an extension proposed to Henseler's (2007) multigroup analysis (hereafter MGA) by Henseler (2012) and Henseler et al. (2009) is applied.

4.4 Findings

4.4.1 Outer Model Assessment

In the first stage of the analyses, the reflective outer model of the two samples from the UK and Romania is evaluated to determine whether the constructs are reliable and distinct from each other. Table 4.1 shows that the loadings of the eight reflective indicators are above the threshold value of 0.70 (Hair et al. 2011, 2012, 2014, 2017;

Henseler et al. 2009), and the lowest composite reliability value is 0.872 and 0.864 for the Romanian sample and the UK sample, respectively. In line with the suggestion of Henseler et al. (2016a), this study also reports Dijkstra-Henseler's rho (ρ_A, Dijkstra and Henseler 2015), which as evidenced in Table 4.1, takes values above 0.70 in both samples. Consequently, both the indicator and construct reliability criteria are met. Further, results suggest a high degree of convergent validity. The lowest average variance extracted (AVE) values are 0.774 (Romania) and 0.761 (the UK), both above the 0.50 threshold level (Hair et al. 2011, 2012, 2014, 2017; Henseler et al. 2009). Moreover, each construct exhibits a high loadings' average and a narrow range between the maximum and minimum loading values (Chin 2010).

Besides convergent validity, reflective measures are also assessed in terms of discriminant validity. For both samples, results confirm that the eight reflective indicators share more variance with their corresponding constructs than with the others (Chin 2010; Henseler et al. 2016a). This indicates that indicators are correctly assigned to the corresponding three constructs and assures that criteria for discriminant validity at an item level are met (Appendix 4.1, Table 4.9). Furthermore, as data in Table 4.2 show, both the Fornell-Larcker Criterion and the Heterotrait-Monotrait Ratio (HTMT) confirm discriminant validity at the construct level. The square root of AVE values between the constructs and their corresponding measures are above the correlations between each two constructs and the three HTMT values are below the cut-off point of 0.90 both in the Romanian and in the UK samples (Henseler et al. 2015).

In the second stage of the analyses, the composite indicators' weights are inspected for their significance. With the exception of OSQ1 with significant absolute importance, all other indicators exhibit significant weights. Further, a closer inspection of the eight composite indicators reveals that one of them (OSQ3) poses multicollinearity problems in the Romanian sample (Appendix 4.1, Table 4.10). When composite indicators are found to be insignificant one should not straightaway decide for their exclusion from further analysis (Diamantopoulos and Winklhofer 2001; Hair et al. 2014; Henseler et al. 2009), given the risk of altering the meaning of the composite construct (Hair et al. 2014, 2017; Henseler et al. 2016b). Nevertheless, if the level of collinearity is high, Henseler et al. (2009) as well as Hair et al. (2014, 2017) argue in favour of indicator exclusion. Therefore, indicator OSQ3 is eliminated from both samples to maintain comparability. This indicator exhibits a variance inflation factor (VIF) value of above the five threshold value (Hair et al. 2014, 2017), and presents an overlap in its meaning with the indicator OSQ1. Consequently, it is appreciated that the meaning attributed to the overall service quality construct is not altered by the exclusion of OSQ3. Table 4.3 illustrates the evaluation of final results for the two composite measures, namely overall service quality (four indicators) and perceived relationship investment (three indicators).

Taken together, data in Tables 4.1, 4.2 and 4.3 offer support for the quality of the outer model in both the Romanian and the UK sample, and ensure that further analyses can be conducted to address the assessment of the inner model.

Table 4.1 Reliability and convergent validity assessment

	Reliability						Convergent validity					
	Indicator reliability		CR		ρ_A		AVE		Loadings' average		Max.–Min. loading value	
Construct indicator	RO	UK	RO	UK	RO	UK	RO	UK	RO	UK	RO	UK
Competence			0.941	0.913	0.921	0.879	0.843	0.778	0.918	0.881	0.062	0.083
C1: […] is a top competitor in its market	0.931	0.897										
C2: As far as I know […] is recognized world-wide	0.880	0.832										
C3: I believe that […] performs at a premium level	0.942	0.915										
Sympathy			0.957	0.933	0.934	0.899	0.880	0.823	0.938	0.907	0.012	0.017
S1: […] is a company I can identify with better than with other companies	0.932	0.911										
S2: […] is a company I regret more if it didn't exist anymore than I would with other companies	0.938	0.897										
S3: I regard […] as a likeable company	0.944	0.913										
Loyalty Intentions			0.872	0.864	0.776	0.725	0.774	0.761	0.879	0.871	0.087	0.073
L1: I intend to use […]'s services in the future too	0.922	0.908										
L2: I am likely to continue my relationship with […] even if fees increase	0.835	0.835										

Notes: *AVE* average variance extracted, *CR* composite reliability, ρ_A rho_A, *RO* Romania, *UK* United Kingdom

Table 4.2 Discriminant validity assessment

	Fornell-Larcker Criterion[a]						The Heterotrait-Monotrait ratio (HTMT)[b]			
	RO			UK			RO		UK	
Construct	Competence	Sympathy	Loyalty intentions	Competence	Sympathy	Loyalty intentions	Loyalty intentions	Competence	Loyalty intentions	Competence
Competence	*0.918*			*0.882*			0.746 [0.661; 0.818]		0.745 [0.656; 0.822]	
Sympathy	0.797	*0.938*		0.695	*0.907*		0.778 [0.687; 0.852]	0.859 [0.800; 0.901]	0.779 [0.681; 0.864]	0.778 [0.695; 0.845]
Loyalty intentions	0.618	0.655	*0.880*	0.590	0.624	*0.872*				

Notes: [a]Diagonal elements (in italics) are the square root of AVE between the constructs and their corresponding measures while the off-diagonal elements are the correlations between each two constructs for each of the two samples

[b]The numbers in brackets are the 95% bias-corrected and accelerated confidence intervals of the HTMT statistic (Henseler et al. 2015) while the numbers outside the brackets represent the HTMT values for each of the two samples

RO Romania, *UK* United Kingdom

Table 4.3 Composite indicators' relevance

	RO				UK			
Construct indicator	VIF	*Weight*	95% BCa confidence interval	Significance?	VIF	*Weight*	95% BCa confidence interval	Significance?
Overall service quality (OSQ)								
OSQ1: […] offers high quality products and services	3.481	0.283	[0.127,0.445]	Yes	3.432	0.165	[0.025,0.295]	Yes
OSQ2: […] is a strong, reliable company	3.327	0.280	[0.129,0.418]	Yes	3.069	0.376	[0.238,0.522]	Yes
OSQ4: […] develops innovative services	2.619	0.343	[0.209,0.468]	Yes	2.412	0.234	[0.123,0.344]	Yes
OSQ5: […] offers services that are a good value for the money	2.509	0.216	[0.084,0.352]	Yes	2.341	0.356	[0.261,0.451]	Yes
Perceived relationship investment (PINV)								
PINV1: […] makes efforts to increase regular customers' loyalty	3.449	0.306	[0.160,0.454]	Yes	3.108	0.176	[0.045,0.301]	Yes
PINV2: […] really cares about keeping regular customers	4.552	0.328	[0.171,0.493]	Yes	4.160	0.370	[0.217,0.533]	Yes
PINV3: I believe […] cares about satisfying my needs	3.298	0.437	[0.292,0.589]	Yes	3.145	0.524	[0.372,0.668]	Yes

Notes: *BCa confidence interval* bias-corrected and accelerated bootstrap confidence interval, *VIF* variance inflation factor, *RO* Romania, *UK* United Kingdom

4.4.2 Inner Model Assessment

Table 4.4 comprises the evaluation criteria considered in the assessment of the inner model for each of the two samples. As evidenced by the low VIF values, redundant information in the predictive constructs considered in the model is not an issue of concern. Following the rule of thumb proposed by Hair et al. (2011) in marketing research, competence and sympathy are appreciated to exhibit moderate to high coefficients of determination in both samples, while the R^2 values of loyalty intentions are weak both in the Romanian and in the UK sample. Perceived relationship investment shows a moderate R^2 value in the Romanian sample and a weak coefficient of determination in the UK sample. Further, the Stone–Geisser Q^2 values for the three reflective endogenous constructs are all positive in both samples. Taken together, these results are indicative of both in-sample and out-of sample prediction (Sarstedt et al. 2014). Across both samples, the path coefficients between the five constructs in the inner model are all positive and significant ($p < 0.001$), in accordance with the hypothesised relationships.

Overall service quality exerts a positive and significant effect on both sympathy and competence, in support of H1a and H1b. Similarly, the higher customers' overall service quality perceptions, the higher their perceived relationship investments made by service providers which provides support for H2. In line with H3a and H3b, perceived relationship investment has a positive and significant effect on sympathy and competence, which in turn leads to more favourable loyalty intentions, thus providing support for H5a and H5b.

Both H4a and H4b assume a mediating effect between overall service quality and sympathy and between overall service quality and competence, respectively, through perceived relationship investment. Recent studies (e.g. Hayes 2013; Nitzl et al. 2016; Zhao et al. 2010) call for a modern approach in the assessment of mediating effects. Contrary to Baron and Kenny's (1986) logic, this state-of-the-art approach claims that the existence of an association between the independent and the dependent variable neither is a precondition of mediation analysis (Hayes 2013) nor is it deemed relevant in establishing whether the mediating effect exists (Zhao et al. 2010). Instead, it is argued that researchers should begin by investigating the significance of the indirect effect (Hayes 2013; Nitzl et al. 2016; Zhao et al. 2010). Moreover, in the context of PLS, Nitzl et al. (2016) maintain that the step-by-step process as proposed by Baron and Kenny (1986) can lead to biases in the path coefficient estimates, which may further alter the true mediating effects. In agreement with the above suggestions, Table 4.5 presents the two indirect effects and the corresponding bias-corrected 95% confidence intervals.

The indirect effects between overall service quality and sympathy and between overall service quality and competence are significant across both samples, thereby supporting H4a and H4b. This, coupled with the significant direct effects (Table 4.4), is indicative of partial mediation. Furthermore, information about the signs of the direct and indirect effects point towards complementary mediation. Essentially, it is only a portion of the effect, which overall service quality has on each of the two

Table 4.4 Inner model quality criteria

	RO				UK			
	VIF	Path coefficients	95% BCa confidence intervals	Significance?	VIF	Path coefficients	95% BCa confidence intervals	Significance?
OSQ → PINV	1.000	0.742	[0.688, 0.781]	Yes	1.000	0.694	[0.631, 0.744]	Yes
OSQ → Competence	2.227	0.415	[0.326, 0.506]	Yes	1.931	0.528	[0.405, 0.619]	Yes
OSQ → Sympathy	2.227	0.498	[0.413, 0.589]	Yes	1.931	0.395	[0.284, 0.487]	Yes
PINV → Competence	2.227	0.463	[0.367, 0.553]	Yes	1.931	0.292	[0.198, 0.399]	Yes
PINV → Sympathy	2.227	0.411	[0.316, 0.501]	Yes	1.931	0.476	[0.382, 0.582]	Yes
Competence → Loyalty intentions	2.741	0.263	[0.147, 0.393]	Yes	1.934	0.304	[0.188, 0.413]	Yes
Sympathy → Loyalty intentions	2.741	0.445	[0.303, 0.565]	Yes	1.934	0.412	[0.291, 0.531]	Yes
	R^2	Q^2			R^2	Q^2		
Competence	0.673	0.557			0.578	0.438		
Sympathy	0.721	0.630			0.643	0.523		
PINV	0.551	–			0.482	–		
Loyalty intentions	0.454	0.340			0.437	0.323		

Notes: *BCa* confidence interval=bias-corrected and accelerated bootstrap confidence interval; the omission distance D should be a number between 5 and 10 so that the number of observation divided by D does not result in an integer (Hair et al. 2014, 2017); in this case, the omission distance was set at 8 (RO(510)/8 = 63.75; UK(525)/8 = 65.63); *VIF* variance inflation factor, *RO* Romania, *UK* United Kingdom

Table 4.5 Mediating effects assessment

	RO				UK			
	Indirect effect	95% BCa confidence intervals	VAF (%)	Significance?	Indirect effect	95% BCa confidence intervals	VAF (%)	Significance?
OSQ → PINV → Competence	0.344	[0.271, 0.415]	45.274	Yes	0.203	[0.137, 0.281]	27.762	Yes
OSQ → PINV → Sympathy	0.305	[0.232, 0.375]	38.010	Yes	0.330	[0.266, 0.404]	45.553	Yes

Notes: *BCa confidence interval* bias-corrected and accelerated bootstrap confidence interval, *VAF* variance accounted for, *RO* Romania, *UK* United Kingdom

Table 4.6 Predictive validity assessment

	RO		UK	
Construct	$R^2_{H\_calculated}$	R^2_T	$R^2_{H\_calculated}$	R^2_T
Investment	0.560	0.555	0.477	0.475
Competence	0.651	0.671	0.595	0.570
Sympathy	0.788	0.686	0.625	0.653
Loyalty intentions	0.595	0.395	0.393	0.456

Notes: $R^2_{H\_calculated}$ calculated coefficients of determination in the holdout samples; R^2_T coefficients of determination in the training sample; *RO* Romania, *UK* United Kingdom

dependent variables, that is mediated through perceived relationship investment. Data in Table 4.5 also show that, for both relationships and across both countries, the mediated effect explains more than 20% and less than 80% of the total effect, thus confirming partial mediation (Hair et al. 2014).

In response to the call made by Woodside (2016), predictive validity analysis is conducted subsequently to further cross-validate the model. In doing so, the approach and guidelines of Cepeda-Carrión et al. (2016) are followed. First, as suggested by Hair et al. (2011), each of the two samples is divided into a training and a holdout sample, with approximately 70% and 30% of the cases, respectively (training sample: Romania = 357 cases, the UK = 368 cases; holdout sample: Romania = 153 cases, the UK = 157 cases). This stage is important as results from the training sub-samples (weights and path coefficients) are used to predict the scores in the holdout sub-samples (Cepeda-Carrión et al. 2016 can be referred to for further details).

As evidenced in Table 4.6, the small differences between calculated R^2 values in the two holdout samples and the corresponding R^2 values in the training samples show that the model has good predictive validity.

4.4.3 Measurement Invariance Assessment

The previous two subsections tested the theoretical framework presented in Fig. 4.1 across both the Romanian and the UK samples. Differences and similarities are easily observable between the results obtained across the two samples. Nevertheless, it is important to establish that they do not occur because of the differences in the content and meaning attributed to the latent variables by respondents in the two groups. Failure to establish this aspect may lead to wrong conclusions when model estimates are compared (Hair et al. 2017; Henseler et al. 2016b; Knoppen et al. 2015; Schlägel and Sarstedt 2016). That is, in cross-national research it is necessary to assess measurement equivalence to prevent erroneous or ambiguous assertions (Steenkamp and Baumgartner 1998). Similarly, Henseler et al. (2016b) and Knoppen et al. (2015) point to the need for ensuring measurement equivalence whenever there is a scope to pool data from different subsamples with the purpose of increasing the generalizability of results.

This study aims to show that relationships between constructs in the theoretical framework proposed here are similar across the two sub-samples and to increase the generalizability of results. To this end, the measures must demonstrate full measurement invariance. Consequently, in line with the guidelines of Henseler et al. (2016b) for PLS applications, and following the illustration provided by Schlägel and Sarstedt (2016), further analyses drawing on the MICOM procedure are performed to assess measurement invariance. Each of the three hierarchically interrelated steps of the MICOM procedure (configural invariance, compositional invariance, and equality of composite mean values and variances) is addressed next.

The issue of configural invariance (MICOM, Step 1) was first addressed during the questionnaire design stage by ensuring that the meaning of indicators measuring specific constructs in the model was retained in the translation process of the questionnaire, which was originally developed in English. To accomplish this, two different persons translated the English version into Romanian by two distinct persons, where one of translators possessed solid knowledge in the marketing field. The resulting two Romanian versions of the questionnaire were compared to resolve discrepancies and the final version was further inspected for meaning equivalence, naturalness in wording, and comprehensibility by two distinct marketing researchers. Finally, the Romanian version of the questionnaire was discussed with two banking services users to ensure that the statements and constructs in the questionnaire are thoroughly understood. Configural invariance was further ensured by identical data treatment and algorithm settings in both samples.

The second step in the MICOM procedure supposes the assessment of equality in composite scores across the two groups. Following the suggestions made by Henseler et al. (2016b), the permutation procedure with 10,000 permutations is run. Data in Table 4.7 reveal that, for each construct in the theoretical framework, the corresponding correlation between the composite scores in the two groups is higher than the 5-percent quantile of the empirical distribution of cu, thus confirming compositional invariance.

Table 4.7 Compositional invariance and equality of composite mean values and variances

	Compositional invariance assessment		Equality of the composite mean values and variances			
Construct	c	5% quantile of c_u	Mean difference (RO–UK)	$CI_{95\%}$	Variance difference (RO–UK)	$CI_{95\%}$
Competence	0.99992	0.99964	0.161	[−0.124, 0.122]	0.308	[−0.205, 0.204]
Sympathy	0.99996	0.99992	0.299	[−0.123, 0.123]	0.287	[−0.189, 0.189]
Loyalty intentions	0.99978	0.99878	0.061	[−0.123, 0.123]	0.264	[−0.190, 0.186]
OSQ	0.99310	0.98595	0.028	[−0.122, 0.122]	0.232	[−0.214, 0.216]
PINV	0.99693	0.98972	0.247	[−0.121, 0.124]	0.174	[−0.175, 0.178]

Notes: $CI_{95\%}$ 95% confidence interval, *RO* Romania, *UK* United Kingdom

Altogether, results obtained in the first two steps of the MICOM procedure provide support that the comparison of path coefficients in the inner model across the two sub-samples does not result in misleading conclusions. Table 4.7 also shows that the 95% confidence intervals of differences in construct means and logarithms of variances between the two sub-samples include zero. Hence, full measurement invariance is established, and data can be pooled to increase the generalizability of results.

4.4.4 Cross-Country Comparison of Causal Relationships

The results of measurement invariance analysis confirm that both configural and compositional invariance is established. Consequently, it is admissible to compare the path coefficients in the inner model across the two samples. Data in Table 4.8 report the results of PLS-MGA proposed by Henseler (2012).

In light of PLS-MGA results, path coefficients estimates may be deemed invariant across the two samples. It is only the difference in the direct effect from perceived relationship investment to competence that is significant ($p < 0.05$).

4.4.5 Pooled Data Model Assessment

Composite mean values and variances across the two samples have been shown to be equal (Table 4.7). Furthermore, most effects in the inner model are invariant across the Romanian and the UK samples. Taken together, measurement invariance analysis and PLS-MGA results support the decision to pool the data. Nevertheless,

Table 4.8 Cross-country comparison of causal relationships

	Path coefficients (RO)	Path coefficients (UK)	Difference in path coefficients (RO–UK)	*p*-value of difference
OSQ → PINV	0.742***	0.694***	0.048	0.095
OSQ → Competence	0.415***	0.528***	−0.112	0.943
OSQ → Sympathy	0.498***	0.395***	0.103	0.062
PINV → Competence	0.463***	0.292***	0.171	0.006
PINV → Sympathy	0.411***	0.476***	−0.065	0.824
Competence → Loyalty intentions	0.263***	0.304***	−0.041	0.685
Sympathy → Loyalty intentions	0.445***	0.412***	0.033	0.355

Notes: ***$p < 0.001$; *RO* Romania, *UK* United Kingdom

before proceeding to interpret results for the aggregated data, it is recommended to run the latent class analysis (Hair et al. 2016, 2017; Ringle et al. 2010; Schlägel and Sarstedt 2016) and to specify an interaction effect to account for the difference in the relationship between perceived relationship investment and competence (Hair et al. 2017; Henseler et al. 2016b).

In line with the suggestions made by Hair et al. (2016, 2017), and Ringle et al. (2010), the latent class analysis was employed to ensure that results for the pooled data are not biased by unobserved heterogeneity. In doing so, the FIMIX-PLS algorithm is preferred, given its applicability and effectiveness, regardless of whether common or composite factors are used (Sarstedt and Ringle 2010). Moreover, the FIMIX-PLS provides the advantage of making it relatively easy to identify the number of latent classes by the examination of several model selection criteria (Sarstedt 2008). Among these, the CAIC criterion was shown to have the highest performance when composite measures are used (Sarstedt et al. 2009). More recently, research has proved that the joint consideration of the AIC_3 and CAIC criteria performs most favourably in pointing towards the appropriate number of segments when the same number of latent classes is indicated by both criteria (Hair et al. 2016; Sarstedt et al. 2011). Other couples of model selection criteria that can be considered in the decision regarding the number of segments are AIC_3 and BIC, AIC_4 and BIC and MDL_5 and AIC (Hair et al. 2016). Apart from these criteria, the normed entropy statistic (EN) should also be inspected to make sure that the FIMIX_PLS segmentation solution results in well-separated segments (Hair et al. 2016; Sarstedt and Ringle 2010; Sarstedt et al. 2011). Another useful indicator that helps in establishing the correct number of segments is the relative size of each group suggested by the FIMIX-PLS algorithm (Sarstedt and Ringle 2010; Hair et al. 2014, 2017).

To check if the aggregated data are affected by unobserved heterogeneity, the FIMIX-PLS algorithm is successively run for one to eight latent classes. Results suggest that there is no heterogeneity. In light of the segment sizes, only a two and three-segment solution are reasonable. Solutions with more than three classes exhibit segments of relative sizes as low as 1.9%. Segments of a small relative size may be deemed unimportant from a managerial perspective, and are therefore of marginal importance for interpretation (Sarstedt and Ringle 2010). Moreover, each segment should meet the minimum sample size requirements (Hair et al. 2014, 2017), a prerequisite that is not met by the remaining five solutions. Further, none of the first three couples of model selection criteria indicates the same number of segments. The MDL_5 points to the two-segment solution. Nevertheless, this criterion is known to underestimate the correct number of segments (Hair et al. 2016). For the three-segment solution that is acceptable in terms of segment sizes, the EN statistic does not reach the minimum threshold of 0.5. Therefore, the segments to be appreciated well-separated (Ringle et al. 2010; Sarstedt and Ringle 2010). Departing from the FIMIX-PLS findings, it can be concluded that there is no critical level of heterogeneity and that data can be pooled for further analyses.

Subsequently, given the lack of structural invariance between perceived relationship investment and competence, the binary variable 'country' (Romania = 0 and the UK = 1) serves as a moderator for the relationship between the two constructs.

Due to the composite nature of the variable perceived relationship investment, the two-stage approach is favoured for the creation of the interaction term. Results show that all the minimum requirements for the outer model are met for both reflective and composite measures (see Appendix 4.2). Data in Table 4.11 (Appendix 4.2) reveal that all indicators exhibit loadings above 0.70. Furthermore, both composite reliability and rho_A take values higher than 0.70. Consequently, reliability criteria are met both at an item and at a construct level. Table 4.11 (Appendix 4.2) further shows that AVE values surpass the threshold of 0.50. This finding, coupled with the high loadings' average and the small difference between the maximum and the minimum loadings' values per construct support the three reflective construct measures' convergent validity. The analysis also confirms discriminant validity. The loadings are compared with the corresponding cross-loadings and the squared loadings are compared with the corresponding squared cross-loadings to assess discriminant validity at an item level (Appendix 4.2, Table 4.12). Discriminant validity at a construct level is established by the comparison of the square root of AVE values with construct correlations as well as by the examination of HTMT values, which are all below the threshold of 0.85 (Appendix 4.2, Table 4.13). Table 4.14 (Appendix 4.2) reveals that each of the two composite measures meets the outer model quality criteria. The VIF Indicator values do not reach the cut-off point of 5 and weights are all significant ($p < 0.001$).

Further analyses aim to test the nine hypotheses corresponding to the relationships in the inner model. Data in Table 4.15 (Appendix 4.2) reveal that all direct effects are positive and significant ($p < 0.001$). Moreover, the effect of overall service quality on competence and sympathy is mediated through perceived relationship investment (Appendix 4.2, Table 4.16). Similar to findings across the two countries, perceived relationship investment only partially mediates the relationship between overall service quality and the two endogenous constructs. Consequently, results for the pooled data sample confirm the hypothesised relationships. The high coefficients of determination and the positive Q^2 values in Table 4.15 (Appendix 4.2) further demonstrate that the inner model has satisfactory in-sample and out-of sample predictive relevance. Results also show that the simple effect of perceived relationship investment on competence is 0.379 and the moderating effect has a significant value of −0.063 ($p < 0.01$). That is, when the mean value of the moderator variable (i.e. 'country') increases by one standard deviation unit, the strength of the relationship between the perceived relationship investment and competence is expected to decrease to 0.316. Further, in line with Kenny's (2016) guidelines, the interaction term's effect size is moderate, thus only modestly contributing to the explanation of the perceived competence ($f^2 = (0.632–0.628)/(1–0.632) = 0.01$).

To test the predictive validity of the model, the overall sample is randomly divided in the training sample (725 observations) and the holdout sample (310 observations). Complying with the guidelines proposed by Cepeda-Carrión et al. (2016), the predicted scores of the endogenous constructs in the holdout sample are computed. The marginal differences between calculated R^2 values in the holdout sample and the corresponding R^2 values in the training sample are indicative of a good predictive validity of the model (Appendix 4.2, Table 4.17).

4.5 Discussion and Conclusions

Corporate reputation is a strategic asset that companies may use to influence stakeholders' behaviour, thereby facilitating their long-term success. A strong reputation is particularly important for companies operating in business contexts that are characterised by long-lasting contractual arrangements. Among these companies, banking institutions need to undertake efforts to ensure that their customers decide to continue the relationship. This study proposed a framework built on inferences rooted in the signalling and social exchange theories. These premises aimed to explain how service quality and relationship investments influence the competence and sympathy of banks as perceived by customers. Finally, this study proposed to demonstrate that the behaviour of bank customers is in accordance with their beliefs and attitudes. As emphasised in the following two subsections, findings reported in this study are relevant from both a theoretical and a managerial perspective.

4.5.1 Research Implications

By addressing the three rationales presented in the introduction, this study makes several contributions to the existing literature. First, this study addresses the call of other researchers (e.g. Terblanche 2014; Walsh et al. 2009a, b), who emphasised that proper consideration of end-use customers is a precondition for the long-term success of companies. Thus, corporate reputation is acknowledged as highly important for long-lasting relationships with end-use customers, one of the most valuable stakeholders, and for survival and success in a competitive environment.

Second, despite the increased interest in the relationship between service quality and corporate reputation, past research has generally adopted the view that corporate reputation is a signal of quality. This study followed the opposite approach and investigated the impact that end-use customers' service quality perceptions exert on their corporate reputation perceptions. Through this approach, this study adheres to the perspective that customers who have used the services of a company have already experienced the actual quality and can therefore make quality judgements. Consequently, building on the signalling theory, it was claimed that quality becomes a signal for corporate reputation. This study's findings confirm that quality is an important determinant of corporate reputation and support the approach proposed by Schwaiger (2004) and later adopted by Eberl (2010), and Sarstedt and Schloderer (2010). Across both samples, results show that overall service quality perceptions directly influence the cognitive and affective dimensions of corporate reputation. In terms of the effects' magnitude, results slightly differ from one sample to another. In accordance with Sarstedt and Schloderer's (2010) findings in a non-profit context, the direct and total effect of overall service quality on sympathy is greater than that on competence in the Romanian sample. Conversely, in the UK sample perceptions about banks' competence are to a greater extent influenced by customers' service quality perceptions, thus confirming Schwaiger's (2004) and Eberl's (2010) findings. Though invariant across samples, these results may indicate that service quality can be a stronger signal of either banks' competence or banks' sympathy among their end-use customers, depending on the environmental context. Considering the uncertainty-avoidance dimension of culture, it may be expected that bank customers from countries with low uncertainty-avoidance scores (such as the UK) feel less threatened by ambiguous situations, more easily accept taking financial risks, supress their emotions (Hofstede et al. 2010), and are less reliant on cues in forming their attitudes (such as service quality signals) (Bartikowski et al. 2011) than the customers in high uncertainty-avoidance countries (such as Romania). Accordingly, service quality perceptions are supposed to exert a weaker impact on sympathy and competence, the two dimensions of reputation, in the UK sample. The results in this study only partially confirm this assumption since bank customers in the UK, compared to those in Romania, place greater importance on quality signals in forming their perceptions of banks' competence. One reason for these findings may be rooted in the consequences of the recent financial crisis. In this respect, Edelman (n.d.) informs that the level of consumers' trust in the UK banks decreased from 42% in 2007 and 2008 to 16% in 2011 and hardly reached 32% in 2015. Contrarily, the Romanian Banking Association (2016) reported that the level of trust in banks of the consumers in Romania increased from 33% in 2011 to 40% in 2016. Consequently, this study findings may point towards a more stringent need of the bank customers in the UK to rely on quality cues provided by their banks when appreciating the cognitive component of reputation.

Third, this study is the first to address empirically the relationship between perceived relationship investment and corporate reputation. Moreover, the relationship between service quality and corporate reputation was hypothesised to be mediated through perceived relationship investment. Drawing on the signalling theory, service quality perceptions were argued to influence customers' perceptions regarding the effort made by their service providers to maintain the relationship. In turn, relying on the norm of reciprocity, which is a key concept in the social exchange theory, positive relationship investment perceptions were expected to significantly contribute to increasing customers' sympathy towards banks and their positive perceptions about the competence of the banks. The hypothesised relationships are confirmed across both samples. This study complements prior research that investigated either the impact of service quality on perceived relationship investment (e.g. De Wulf et al. 2003; Liang and Wang 2006) or the effect of service quality on corporate reputation (e.g. Eberl 2010; Sarstedt and Schloderer 2010; Schwaiger 2004). Findings from this study demonstrate that the overall service quality can simultaneously serve as a signal of banks' unobservable efforts to maintain the relationship with their customers and of their reputation. Further, extending prior research, this study shows that banks' efforts to develop and maintain long-term relationships with their end-use customers lead to a favourable feedback, as expressed by enhanced corporate reputation perceptions. Nevertheless, results concerning the relative importance of perceived relationship investment in explaining the affective and the cognitive components of reputation differ across the two samples. While in the UK sample sympathy is to a greater extent influenced by perceived relationship investment than in the Romanian sample, the opposite is true for the relationship between perceived relationship investment and competence. These findings may be due to the cultural differences between the UK and Romania with respect to both individualism-collectivism and the power distance cultural dimensions. People in high individualist and low power distance societies (such as the UK) expect that everybody is treated alike and preferential treatment or privileges are not valued. Contrarily, such preferential treatment and privileges are considered unethical. Conversely, in collectivist and high power distance societies (such as Romania), preferential treatment, privileges, and favouritism for 'our group' are not proof of an unethical behaviour, but may be deemed normal and even expected (Hofstede et al. 2010). In this study, variables measuring perceived relationship investment are rather worded so as to reflect banks' efforts to maintain and develop favourable relationships with all their customers, regardless of their status. Consequently, it is not surprising that, compared to the customers in Romania, the UK bank customers' sympathy towards their banks is to a greater extent influenced by perceived relationship investment. Likewise, a higher effect of perceived relationship investment on competence in the Romanian sample can be easily anticipated since customers do not get the feeling that, compared to others, they are treated better in the relationship they have with their bank or service providers. Moreover, findings in this study extend prior research by demonstrating the role of perceived relationship investment as a mediator between overall service quality and corporate reputation.

Fourth, building on the social exchange theory, this study hypothesised that favourable corporate reputation perceptions are a determinant of relationship loyalty. Contrary to Eberl's (2010) conceptualisation in the telecommunication industry, that only the affective dimension of reputation exerts a positive impact on loyalty, the framework proposed in this study emphasises that the cognitive dimension is equally important in fostering customers' intentions. Across both samples, perceived competence and customers' sympathy towards their banks are shown to have a positive impact on loyalty intentions. These findings confirm prior research (e.g. Bartikowski and Walsh 2011; Walsh et al. 2009a, 2014) and support the idea that consumers act in accordance with their beliefs and attitudes (Walsh et al. 2014). Moreover, findings are in agreement with the contention that a strong reputation influences customers' decisions about a company's offering (Alniacik et al. 2012; Vidaver-Cohen 2007). Consequently, it is argued that corporate reputation perceptions leading to relationship loyalty embody consumers' reciprocation to their service providers' relationship investments, thereby reflecting customers' positive attitude towards maintaining the relationship.

Finally, as full measurement invariance was established, the data from the two samples was pooled. Analyses conducted on the aggregated data confirm all the nine hypothesised relationships, thus increasing the generalizability of results. The research framework proposed and validated in this study demonstrates that banks' efforts to improve the quality of services are a signal of both their endeavour to maintain the relationship with end-use customers and their reputation. In addition, it shows that banks' efforts to develop and maintain long-term relationships with customers lead to favourable corporate reputation perceptions. These, in turn, ultimately determine consumers' decision to reciprocate by sending back a positive feedback, as expressed through the predisposition to maintain the relationship.

4.5.2 Managerial Implications

Perhaps the most important finding for bank managers relates to the return on their investment to increase the quality of banking services and maintain long-lasting relationships with their end-use customers. Results show that the overall service quality is an important determinant of both perceived competence of banks and the sympathy that customers develop towards their financial services providers. Similarly, investments in the quality of services lead to favourable consumer perceptions regarding the efforts made by banks to maintain the relationship. These findings are of high importance as they emphasise that end-use bank customers recognize that service quality acts as a signal for both relationship investment and reputation. Consequently, banking institutions that embrace the role of signallers should consider service quality as a distinguishing characteristic that would help them inform customers about the unobservable efforts undertaken to maintain the relationship and build a strong reputation. In light of these findings, bank managers should 'proceed first from perceived service quality to relationship establishment' (Liang and Wang 2006, p. 137). Specifically, it is important that banks offer high quality and innovative services that are perceived to have a good value for money. These would stand as proof of banks' reliability and would convey their interest in satisfying end-use customers' needs, retaining regular customers, and increasing customer loyalty.

Results also demonstrate that along with overall service quality, perceived relationship investment is essential for building a strong reputation. Both the perceived competence and customers' sympathy towards their banking institutions are largely dependent on the efforts undertaken by banks to improve the quality of their offering and maintain the relationship with their customers. This is an important finding because corporate reputation is an intangible asset (Shamma 2012; Vidaver-Cohen 2007) that can help companies differentiate themselves from competitors (Alniacik et al. 2012). Essentially, this study offers guidance on becoming a likeable bank and a top competitor in the market. Improving the quality of the offering is a route to fostering favourable perceptions of banks' investments in maintaining the relationship with their end-use customers and eventually improving reputation.

Nevertheless, banks conducting business in different markets are also advised to consider the specific context in which corporate reputation needs to be built and monitored. Findings in the Romanian sample provide evidence that service quality cues have a greater contribution to fostering customers' sympathy towards their banks, while perceived competence of banks is mainly determined by perceived relationship investment. Conversely, in the UK sample, service quality is the key driver of competence, whereas banks' efforts to maintain the relationship with their customers have a greater positive impact on the affective than on the cognitive dimension of reputation. Therefore, managers should make decisions in accordance with banks' objectives in the specific market in which they operate. If the main purpose is to increase customers' perceptions of bank competence, then the banks in Romania and the UK should follow different routes to achieve their objective.

Romanian bank managers are advised to devote resources to actions that communicate their interest in building and maintaining long-lasting relationships with customers. In order to send such messages to customers effectively, bank strategies need to be designed in a manner that it develops customers' confidence in the ability of banks to respect their promises on a long-term basis. The long-lasting contractual arrangements between the two parties, and technological advancements, allow banks to know their customers better. Consequently, making use of customer databases, bank employees can access accurate and up-to-date information that may help them more easily anticipate and address customers' financial needs. Banks can also monitor their customers' behaviour and develop reward and loyalty programs that communicate their interest in maintaining long-term and fruitful relationships with customers. Technology also enables banks to establish and maintain a personalized dialogue with customers and offers support whenever needed. Such efforts undertaken by banks are expected to strengthen customers' beliefs that their financial service provider cares about satisfying and retaining them.

Conversely, when the purpose is to increase customers' perceptions about the competence of the UK banks, managers should make efforts to ensure that service quality cues are visible and efficient enough to instil a belief that the bank is competent. In light of the recent financial crisis, it is important for banks and their employees to be considered as partners of customers and not as lenders. That is, they should become an advisor in order to respond to customers' needs and address their expectations in a better manner. In this respect, it is important that bank managers strive that their employees have good knowledge, expertise and competence to respond to customers' financial requests. Providing services as promised, without delays, and showing interest in promptly and properly handling customers' problems are other examples of actions that signal banks' ability to offer high-quality services to their customers, and increase customers' perceptions of banks' reliability. Furthermore, providing a diverse range of services and, investing in recent innovations that can help customers better fulfil their financial needs should be priorities for banks that aim to be competitive. Considered together, the implementation of such actions would help banks in the UK signal that they are responsible and reliable, which, in turn, would lead to an increase in customers' perceptions of their competence.

This study further confirms the benefits associated with a strong reputation. Researchers note that favourable reputation perceptions lead to customers' intention to continue to purchase the company's products and/or services (Alniacik et al. 2012) as well as to increased loyalty towards the company (e.g. Chun 2005; Walsh and Beatty 2007; Walsh et al. 2014). Indeed, results in this study show that both perceived competence and consumers' sympathy towards their financial providers have a positive effect on relationship loyalty. From a managerial perspective, this finding suggests that banks should carefully monitor their customers' reputation perceptions. Failure to do this may result in customers' defection. Across both samples, sympathy is the main driver of customers' intention to continue the relationship with their banks. This means that banks in Romania, and in the UK, should primarily adopt strategies that are effective in increasing customers' sympathy towards their banks. Since both service quality and relationship investment signals have been found to be significant drivers of sympathy, banks in Romania and the UK should consider investing in issues related to both these antecedents of the affective component of reputation. Although invariant across the two samples, the contribution that each antecedent has in boosting customers' sympathy slightly differs in the two samples. That is, bank managers in Romania should first invest in quality cues, whereas the UK banks' strategies should mainly be designed to facilitate an increase in customers' perceptions that their banks' efforts to maintain the relationship with them are credible. Though sympathy is found to exert a greater impact on customers' intention to maintain the relationship, banks should also give careful attention to the cognitive reputational dimension. Favourable reputation perceptions, including competence perceptions, reap benefits such as customers' intention to use the services in the future and their willingness to continue the relationship with the bank, despite the possibility of a surge in fees.

The framework proposed in this study holds true across samples from both countries, and is confirmed on the aggregated data as well. From a practical perspective, these findings are particularly important for banks conducting business on different markets. Improvements in the quality of services are a signal of relationship investments made by banks and of their reputation. Moreover, perceived relationship investment is a causal factor of corporate reputation in its own right. When banks make efforts to offer high-quality services and maintain the relationship with customers, they send a signal about their unobservable competence and nurture sympathy among end-use consumers. Bank competence and increased customer sympathy lead to a competitive advantage in that customers want to reciprocate by declaring their intention to maintain the relationship. This compliance between customers' beliefs, attitudes, and behaviour suggests that banks' endeavour to show interest in their end-use customers would lead to benefits in the long run.

4.5.3 Limitations and Further Research Directions

This study fills a gap in the literature by emphasising that service quality along with the more long-term construct of perceived relationship investment are important signals of banks' reputation. Nevertheless, future research may attempt to extend the framework proposed in this study. In this respect, future studies should extend previous findings (e.g. Eberl 2010; Sarstedt and Schloderer 2010; Schwaiger 2004) by investigating the mediating role of perceived relationship investment between other constructs (e.g. perceived performance and, perceived responsibility) and corporate reputation.

This study demonstrates that consumers' behaviour measured through relationship loyalty is in accordance with their reputation perceptions. Future research should extend the framework and establish the relationship between both the cognitive and affective reputational dimensions and other behavioural variables (e.g. citizenship behaviour, cross-buying, share of wallet, and word-of-mouth). Such variables, though proven to be consequences of overall corporate reputation (e.g. in studies of Bartikowski and Walsh 2011; Jeng 2008; Shamma and Hassan 2009; Walsh et al. 2009a, 2014), have not yet been considered in studies that regard reputation a two-dimensional construct.

The results in this study extend previous findings by showing that perceived relationship investment partially mediates the relationship between overall service quality and the two dimensions of corporate reputation. Moreover, results point to complementary mediation. These findings offer promising avenues for research. Zhao et al. (2010) argue that complementary mediation stands as a proof of a direct effect that can be further explained by additional mediators, which are omitted from the model. Though only partially confirmed, the framework proposed by Wang et al. (2006) hypothesised that service quality positively influences brand equity, which in turn was assumed as a precursor of corporate reputation. Satisfaction is a construct that proved to be a determinant of corporate reputation (e.g. Bontis et al. 2007; Walsh et al. 2009a) and an important mediator in the relationship between service quality and reputation (e.g. Carmeli and Tishler 2005). Therefore, future research could test whether brand equity and/or satisfaction have a mediating role, along with perceived relationship investment, between overall service quality and corporate reputation.

This study uses data collected from the end-use consumers in two European countries. Researchers are encouraged to test the hypotheses proposed in this study in different cultural contexts. With one exception only, all path coefficients estimates are invariant across the Romanian and the UK samples. Despite this finding, results may slightly differ if the framework is to be tested in other countries. Therefore, further validation of the model and generalization of results would be essential for international banks. Such an approach would answer the call of Franke and Richey (2010) who argue that a minimum of seven countries are recommended for comparison for enabling researchers to make faithful international generalizations.

Finally, researchers are encouraged to test the relationships proposed in this study in other service areas as well. Gardberg (2006) argues that the industry in which a company operates may influence its reputation, and thus the relationship between this construct and its antecedents and consequences. Consequently, such an attempt would be of great significance from a theoretical perspective. Findings would extend academics' understanding of how signalling and social exchange theories conjunctively explain the relationships between service quality, perceived relationship investment, corporate reputation and the reciprocal behaviour of customers in different service settings.

Appendix 4.1

Table 4.9 Discriminant validity at an item level

	RO						UK					
	Loadings and cross-loadings			Squared loadings and cross-loadings			Loadings and cross-loadings			Squared loadings and cross-loadings		
Indicator	Competence	Sympathy	Loyalty intentions	Competence	Sympathy	Loyalty intentions	Competence	Sympathy	Loyalty intentions	Competence	Sympathy	Loyalty intentions
C1	0.931	0.773	0.611	0.867	0.598	0.374	0.897	0.651	0.558	0.805	0.424	0.312
C2	0.880	0.627	0.483	0.774	0.393	0.233	0.831	0.470	0.422	0.691	0.221	0.178
C3	0.942	0.780	0.595	0.888	0.608	0.354	0.915	0.687	0.563	0.836	0.472	0.317
S1	0.731	0.932	0.602	0.535	0.869	0.362	0.630	0.911	0.554	0.397	0.830	0.307
S2	0.719	0.938	0.597	0.516	0.879	0.357	0.585	0.896	0.514	0.342	0.804	0.264
S3	0.790	0.944	0.642	0.624	0.892	0.412	0.669	0.913	0.620	0.448	0.834	0.385
LI1	0.610	0.669	0.922	0.372	0.448	0.850	0.583	0.605	0.908	0.340	0.366	0.824
LI2	0.459	0.454	0.835	0.211	0.206	0.697	0.431	0.470	0.835	0.186	0.221	0.697

Notes: *RO* Romania, *UK* United Kingdom

Table 4.10 Composite indicators'relevance

	RO				UK			
Construct indicator	VIF	*Weight*	95% BCa confidence interval	Significance?	VIF	*Weight*	95% BCa confidence interval	Significance?
Overall service quality (OSQ)								
OSQ1: […] offers high quality products and services	4.539	0.159	[−0.005, 0.340]	No	3.770	0.059	[−0.094, 0.183]	No
OSQ2: […] is a strong, reliable company	3.979	0.182	[0.032, 0.321]	Yes	3.404	0.267	[0.139, 0.405]	Yes
OSQ3: […] stands behind the services that it offers	6.976	0.314	[0.139, 0.476]		3.264	0.319	[0.203, 0.427]	
OSQ4: […] develops innovative services	2.799	0.290	[0.167, 0.415]	Yes	2.503	0.177	[0.062, 0.284]	Yes
OSQ5: […] offers services that are a good value for the money	2.715	0.161	[0.022, 0.295]	Yes	2.387	0.314	[0.226, 0.409]	Yes
Perceived relationship investment (PINV)								
PINV1: […] makes efforts to increase regular customers' loyalty	3.449	0.303	[0.152, 0.448]	Yes	3.108	0.178	[0.048, 0.300]	Yes
PINV2: […] really cares about keeping regular customers	4.552	0.327	[0.164, 0.490]	Yes	4.160	0.366	[0.216, 0.517]	Yes
PINV3: I believe […] cares about satisfying my needs	3.298	0.441	[0.296, 0.593]	Yes	3.145	0.525	[0.377, 0.662]	Yes

Notes: *BCa confidence interval* bias-corrected and accelerated bootstrap confidence interval, *VIF* variance inflation factor, *RO* Romania, *UK* United Kingdom

Appendix 4.2

Table 4.11 Reliability and convergent validity assessment

	Reliability			Convergent validity		
Construct indicator	Indicator reliability	CR	ρ_A	AVE	Loadings' average	Max.–Min. loading value
Competence		0.929	0.904	0.813	0.901	0.074
C1: […] is a top competitor in its market	0.917					
C2: As far as I know […] is recognized world-wide	0.856					
C3: I believe that […] performs at a premium level	0.930					
Sympathy		0.946	0.919	0.855	0.924	0.008
S1: […] is a company I can identify with better than with other companies	0.924					
S2: […] is a company I regret more if it didn't exist anymore than I would with other companies	0.921					
S3: I regard […] as a likeable company	0.929					
Loyalty intentions		0.868	0.753	0.768	0.875	0.082
L1: I intend to use […]'s services in the future too	0.916					
L2: I am likely to continue my relationship with […] even if fees increase	0.834					

Notes: *AVE* average variance extracted, *CR* composite reliability, ρ_A rho_A

Table 4.12 Discriminant validity at an item level

	Loadings and cross-loadings			Squared loadings and cross-loadings		
Indicator	Competence	Sympathy	Loyalty intentions	Competence	Sympathy	Loyalty intentions
C1	0.917	0.721	0.589	0.841	0.520	0.346
C2	0.856	0.551	0.455	0.733	0.304	0.207
C3	0.930	0.742	0.581	0.865	0.551	0.337
S1	0.686	0.924	0.576	0.471	0.853	0.332
S2	0.661	0.921	0.556	0.437	0.848	0.310
S3	0.738	0.929	0.632	0.544	0.863	0.400
LI1	0.599	0.639	0.916	0.359	0.408	0.839
LI2	0.446	0.458	0.834	0.199	0.210	0.696

Table 4.13 Discriminant validity assessment

	Fornell-Larcker criterion[a]			The Heterotrait-Monotrait ratio (HTMT)[b]	
Construct	Competence	Sympathy	Loyalty intentions	Loyalty intentions	Competence
Competence	*0.902*			0.746 [0.685; 0.799]	
Sympathy	0.753	*0.925*		0.775 [0.711; 0.835]	0.825 [0.779; 0.864]
Loyalty intentions	0.606	0.638	*0.876*		

Notes: [a]Diagonal elements (in italics) are the square root of AVE between the constructs and their corresponding measures while the off-diagonal elements are the correlations between each two constructs for each of the two samples

[b]The numbers in brackets are the 95% bias-corrected and accelerated confidence intervals of the HTMT statistic (Henseler et al. 2015) while the numbers outside the brackets represent the HTMT values for each of the two samples

Table 4.14 Composite indicators'relevance

Construct indicator	VIF	Weight	95% BCa confidence interval	Significance?
Overall service quality (OSQ)				
OSQ1: […] offers high quality products and services	3.439	0.225	[0.115,0.338]	Yes
OSQ2: […] is a strong, reliable company	3.164	0.341	[0.230,0.438]	Yes
OSQ4: […] develops innovative services	2.518	0.297	[0.208,0.388]	Yes
OSQ5: […] offers services that are a good value for the money	2.414	0.265	[0.175,0.353]	Yes
Perceived relationship investment (PINV)				
PINV1: […] makes efforts to increase regular customers' loyalty	3.314	0.244	[0.149,0.338]	Yes
PINV2: […] really cares about keeping regular customers	4.416	0.346	[0.228,0.459]	Yes
PINV3: I believe […] cares about satisfying my needs	3.240	0.481	[0.381,0.586]	Yes

Notes: *BCa confidence interval* bias-corrected and accelerated bootstrap confidence interval, *VIF* variance inflation factor

Table 4.15 Inner model quality criteria

	Without interaction effect				With interaction effect			
	VIF	Path coefficients	95% BCa confidence interval	Significance?	VIF	Path coefficients	95% BCa confidence interval	Significance?
OSQ → PINV	1.000	0.713	[0.672; 0.746]	Yes	1.000	0.713	[0.673; 0.745]	Yes
OSQ → Competence	2.033	0.469	[0.396; 0.535]	Yes	2.063	0.468	[0.397; 0.536]	Yes
OSQ → Sympathy	2.033	0.430	[0.365; 0.495]	Yes	2.033	0.430	[0.366; 0.499]	Yes
PINV → Competence	2.033	0.387	[0.319; 0.457]	Yes	2.089	0.379	[0.309; 0.447]	Yes
PINV → Sympathy	2.033	0.465	[0.395; 0.530]	Yes	2.033	0.465	[0.396; 0.531]	Yes
Competence → Loyalty intentions	2.313	0.290	[0.205; 0.372]	Yes	2.313	0.290	[0.212; 0.376]	Yes
Sympathy → Loyalty Intentions	2.313	0.420	[0.330; 0.507]	Yes	2.313	0.420	[0.325; 0.501]	Yes
Country → Competence	–	–	–	–	1.027	−0.027	[−0.065; 0.012]	No
Country*PINV → Competence	–	–	–	–	1.011	−0.063	[−0.113; −0.017]	Yes
	R^2	Q^2			R^2	Q^2		
Competence	0.628	0.502			0.632	0.504		
Sympathy	0.685	0.582			0.685	0.582		
PINV	0.508	–			0.508	–		
Loyalty intentions	0.444	0.332			0.444	0.332		

Notes: *BCa confidence interval* bias-corrected and accelerated bootstrap confidence interval, the omission distance D should be a number between 5 and 10 so that the number of observation divided by D does not result in an integer (Hair et al. 2014, 2017); in this case, the omission distance was set at 8 (1035/8 = 128.375), *VIF* variance inflation factor

Table 4.16 Mediating effects assessment

	Without interaction effect				With interaction effect			
	Indirect effect	95% BCa confidence interval	VAF (%)	Significance?	Indirect effect	95% BCa confidence interval	VAF (%)	Significance?
OSQ → PINV → Competence	0.276	[0.227, 0.329]	37.02	Yes	0.270	[0.218, 0.319]	36.58	Yes
OSQ → PINV → Sympathy	0.331	[0.281, 0.379]	43.52	Yes	0.331	[0.281, 0.379]	43.52	Yes

Notes: *BCa confidence interval* bias-corrected and accelerated bootstrap confidence interval, *VAF* variance accounted for

Table 4.17 Predictive validity assessment

Construct	$R^2_{H\_calculated}$	R^2_T
Investment	0.496	0.507
Competence	0.577	0.654
Sympathy	0.699	0.682
Loyalty intentions	0.423	0.452

Notes: $R^2_{H\_calculated}$ calculated coefficients of determination in the holdout samples, R^2_T coefficients of determination in the training sample

References

Alniacik, E., Alniacik, U., & Erdogmus, N. (2012). How do the dimensions of corporate reputation affect employment intentions? *Corporate Reputation Review, 15*(1), 3–19.

Barisitz, S. (2004). *The transformation of the Romanian financial and banking sector* (Financial Stability Report No. 7). Vienna: Oesterreichische Nationalbank, 88–100.

Baron, R. M., & Kenny, D. A. (1986). The moderator-mediator variable distinction in social psychological research: Conceptual, strategic, and statistical considerations. *Journal of Personality and Social Psychology, 51*(6), 1173–1182.

Bartikowski, B., & Walsh, G. (2011). Investigating mediators between corporate reputation and customer citizenship behaviors. *Journal of Business Research, 64*(1), 39–44.

Bartikowski, B., Walsh, G., & Beatty, S. E. (2011). Culture and age as moderators in the corporate reputation and loyalty relationship. *Journal of Business Research, 64*(9), 966–972.

Bendapudi, N., & Berry, L. L. (1997). Customers' motivations for maintaining relationships with service providers. *Journal of Retailing, 73*(1), 15–37.

Blau, P. M. (1964). *Exchange & power in social life*. New York: Wiley.

Bontis, N., Booker, L. D., & Serenko, A. (2007). The mediating effect of organizational reputation on customer loyalty and service recommendation in the banking industry. *Management Decision, 45*(9), 1426–1445.

Boulding, W., & Kirmani, A. (1993). A consumer–side experimental examination of signaling theory: Do consumers perceive warranties as signals of quality? *Journal of Consumer Research, 20*(1), 111–123.

Braggion, F., & Ongena, S. (2013). *A century of firm—bank relationships: Did banking sector deregulation spur firms to add banks and borrow more?*. CEPR Discussion Paper No. DP9695. Retrieved October 25, 2016, from https://ssrn.com/abstract=2343705.

Carmeli, A., & Tishler, A. (2005). Perceived organizational reputation and organizational performance: An empirical investigation of industrial enterprises. *Corporate Reputation Review, 8*(1), 13–30.

Cepeda-Carrión, G., Henseler, J., Ringle, C. M., & Roldán, J. L. (2016). Prediction-oriented modeling in business research by means of PLS path modeling: Introduction to a JBR special section. *Journal of Business Research, 69*(10), 4545–4551.

Chibucos, T. R., Leites, R. W., & Weis, D. L. (2005). *Readings in family theory*. Thousand Oaks: Sage.

Chin, W. W. (2010). How to write up and report PLS analyses. In V. E. Vinzi, W. W. Chin, J. Henseler, & H. Wang (Eds.), *Handbook of partial least squares* (pp. 655–690). Heidelberg: Springer.

Chun, R. (2005). Corporate reputation: Meaning and measurement. *International Journal of Management Reviews, 7*(2), 91–109.

Colgate, M., & Smith, B. (2005). Relationships and the internet: The mediating role of a relationship banker. *Journal of Financial Services Marketing, 10*(2), 140–151.

Connelly, B. L., Certo, S. T., Ireland, R. D., & Reutzel, C. R. (2011). Signaling theory: A review and assessment. *Journal of Management, 37*(1), 39–67.

Cropanzano, R., & Mitchell, M. S. (2005). Social exchange theory: An interdisciplinary review. *Journal of Management, 31*(6), 874–900.

Davies, R., Richardson, P., Katinaite, V., & Manning, M. (2010). Evolution of the UK banking system. *Bank of England Quarterly Bulletin, Q4*, 321–332.

Dawar, N., & Parker, P. (1994). Marketing universals: Consumers' use of brand name, price, physical appearance, and retailer reputation as signals of product quality. *Journal of Marketing, 58*(2), 81–95.

De Wulf, K., Odekerken-Schröder, G., & Iacobucci, D. (2001). Investments in consumer relationships: A cross-country and cross-industry exploration. *Journal of Marketing, 65*(4), 33–50.

De Wulf, K., Odekerken-Schröder, G., & Van Kenhove, P. (2003). Investments in consumer relationships: A critical reassessment and model extension. *International Review of Retail, Distribution and Consumer Research, 13*(3), 245–261.

Diamantopoulos, A., & Winklhofer, H. M. (2001). Index construction with formative indicators: An alternative to scale development. *Journal of Marketing Research, 38*(2), 269–277.

Dijkstra, T. K., & Henseler, J. (2015). Consistent partial least squares path modeling. *MIS Quarterly, 39*(2), 297–316.

Dimitriadis, S. (2010). Testing perceived relational benefits as satisfaction and behavioral outcomes drivers. *International Journal of Bank Marketing, 28*(4), 297–313.

Doltu, C. (2000). The evolution of the banking system in Romania. Romanian Center for Economic Policies no. 8, 1–13.

Dunham, B. (2011). The role for signaling theory and receiver psychology in marketing. In G. Saad (Ed.), *Evolutionary psychology in the business science* (pp. 225–256). Heidelberg: Springer.

Eberl, M. (2010). An application of PLS in multi-group analysis: The need for differentiated corporate-level marketing in the mobile communications industry. In V. E. Vinzi, W. W. Chin, J. Henseler, & H. Wang (Eds.), *Handbook of partial least squares* (pp. 487–514). Heidelberg: Springer.

Edelman. (n.d.). Level of trust in banks in the United Kingdom between 2007 and 2015. In Statista—The Statistics Portal. Retrieved March 20, 2017, from https://www.statista.com/statistics/514244/bank-trust-levels-in-uk.

Eurostat. (n.d.). Online banking penetration in Romania from 2006 to 2016. In Statista—the statistics portal. Retrieved March 15, 2017, from https://www.statista.com/statistics/380874/online-banking-penetration-in-romania/.

EZB. (n.d.). Total number of payment cards issued in Romania from 2009 to 2016 (in 1,000s). In Statista—The statistics portal. Retrieved March 15, 2017, from https://www.statista.com/statistics/444436/number-of-payment-cards-in-romania/.

Falkenreck, C., & Wagner, R. (2010). Impact of direct marketing activities on company reputation transfer success: Empirical evidence from five different cultures. *Corporate Reputation Review, 13*(1), 20–37.

Feldman, P. M., Bahamonde, R. A., & Velasquez Bellido, I. (2014). A new approach for measuring corporate reputation. *Revista de Administração de Empresas, 54*(1), 53–66.

Franke, G. R., & Richey, R. G., Jr. (2010). Improving generalizations from multi-country comparisons in international business research. *Journal of International Business Studies, 41*(8), 1275–1293.

Freedom House. (2015a). *Freedom in the World*. Retrieved March 15, 2017, from https://freedomhouse.org/report/freedom-world/2015/united-kingdom and https://freedomhouse.org/report/freedom-world/2015/romania.

Freedom House. (2015b). *Freedom of the press*. Retrieved March 15, 2017, from https://freedomhouse.org/report/freedom-press/2015/united-kingdom and https://freedomhouse.org/report/freedom-press/2015/romania.

Gardberg, N. A. (2006). Reputatie, reputation, réputation, Reputazione, Ruf: A cross-cultural qualitative analysis of construct and instrument equivalence. *Corporate Reputation Review, 9*(1), 39–61.

Hair, J. F., Ringle, C. M., & Sarstedt, M. (2011). PLS-SEM: Indeed a silver bullet. *Journal of Marketing Theory and Practice, 19*(2), 139–151.

Hair, J. F., Sarstedt, M., Ringle, C. M., & Mena, J. A. (2012). An assessment of the use of partial least squares structural equation modeling in marketing research. *Journal of the Academy of Marketing Science, 40*(3), 414–433.

Hair, J. F., Hult, G. T. M., Ringle, C. M., & Sarstedt, M. (2014). *A primer on partial least squares structural equation modeling* (1st ed.). Thousand Oaks: Sage.

Hair, J. F., Sarstedt, M., Matthews, L. M., & Ringle, C. M. (2016). Identifying and treating unobserved heterogeneity with FIMIX-PLS: Part I—method. *European Business Review, 28*(1), 63–76.

Hair, J. F., Hult, G. T. M., Ringle, C. M., & Sarstedt, M. (2017). *A primer on partial least squares structural equation modeling* (2nd ed.). Thousand Oaks: Sage.

Hayes, A. F. (2013). *Introduction to mediation, moderation, and conditional process analysis: A regression-based approach.* New York: Guilford.

Henseler, J. (2007). A new and simple approach to multi-group analysis in partial least squares path modeling. In H. Martens, T. Næs, & M. Martens (Eds.), *Causalities explored by indirect observation proceedings of the 5th International Symposium on PLS and related methods (PLS'07)* (pp. 104–107). Norway: Matforsk, As.

Henseler, J. (2012). PLS-MGA: A non-parametric approach to partial least squares-based multi-group analysis. In W. A. Gaul, A. Geyer-Schulz, L. Schmidt-Thieme, & J. Kunze (Eds.), *Challenges at the interface of data analysis, computer science, and optimization (studies in classification, data analysis, and knowledge organization)* (pp. 495–501). Heidelberg: Springer.

Henseler, J., Ringle, C. M., & Sinkovics, R. R. (2009). The use of partial least squares path modeling in international marketing. In R. R. Sinkovics & P. N. Ghauri (Eds.), *New challenges to international marketing: advances in international marketing* (pp. 277–319). Bingley: Emerald Group.

Henseler, J., Ringle, C. M., & Sarstedt, M. (2015). A new criterion for assessing discriminant validity in variance-based structural equation modelling. *Journal of the Academy of Marketing Science, 43*(1), 1–21.

Henseler, J., Hubona, G., & Ray, P. (2016a). Using PLS path modeling in new technology research: Updated guidelines. *Industrial Management & Data Systems, 116*(1), 2–20.

Henseler, J., Ringle, C. M., & Sarstedt, M. (2016b). Testing measurement invariance of composites using partial least squares. *International Marketing Review, 33*(3), 405–431.

Hofstede, G., Hofstede, G. J., & Minkov, M. (2010). *Cultures and organizations—software in the mind* (3rd ed.). New York: McGraw Hill.

International Monetary Fund. (2013). *Financial access survey.* Retrieved March 15, 2017, from http://data.imf.org/?sk=E5DCAB7E-A5CA-4892-A6EA-598B5463A34C.

Jarvenpaa, S. L., Tractinsky, N., & Saarinen, L. (1999). Consumer trust in an internet store: A cross-cultural validation. *Journal of Computer-Mediated Communication, 5*(2), 45–71.

Jeng, S. P. (2008). Effects of corporate reputations, relationships and competing suppliers' marketing programmes on customers' cross-buying intentions. *The Service Industries Journal, 28*(1), 15–26.

Jeng, S. P. (2011). The effect of corporate reputations on customer perceptions and cross-buying intentions. *The Service Industries Journal, 31*(6), 851–862.

Jin, B., Yong Park, J., & Kim, J. (2008). Cross-cultural examination of the relationships among firm reputation, e-satisfaction, e-trust, and e-loyalty. *International Marketing Review, 25*(3), 324–337.

Jing, L., & Lei, W. (2007). A study on the signaling mechanism of corporate reputation. In *Proceedings of the wireless communications, networking and mobile computing, international conference* (pp. 3277–3280). Shanghai: IEEE.

Keh, H. T., & Xie, Y. (2009). Corporate reputation and customer behavioral intentions: The roles of trust, identification and commitment. *Industrial Marketing Management, 38*(7), 732–742.

Kenny, D.A. (2016). *Moderation.* Retrieved July 28, 2016, from http://davidakenny.net/cm/moderation.htm#WWF.

Kirmani, A., & Rao, A. R. (2000). No pain, no gain: a critical review of the literature on signaling unobservable product quality. *Journal of Marketing, 64*(2), 66–79.

Knoppen, D., Ateş, M. A., Brandon-Jones, A., Luzzini, D., Van Raaij, E., & Wynstra, F. (2015). A comprehensive assessment of measurement equivalence in operations management. *International Journal of Production Research, 53*(1), 166–182.

Liang, C. J., & Wang, W. H. (2006). The behavioural sequence of the financial services industry in Taiwan: Service quality, relationship quality and behavioural loyalty. *The Service Industries Journal, 26*(2), 119–145.

Liao, L. F. (2008). Knowledge-sharing in R&D departments: A social power and social exchange theory perspective. *The International Journal of Human Resource Management, 19*(10), 1881–1895.

Loureiro, S. M. C., & Kastenholz, E. (2011). Corporate reputation, satisfaction, delight, and loyalty towards rural lodging units in Portugal. *International Journal of Hospitality Management, 30*(3), 575–583.

MacKenzie, S. B. (2003). The dangers of poor construct conceptualization. *Journal of the Academy of Marketing Science, 31*(3), 323–326.

Nelson, P. (1970). Information and consumer behavior. *Journal of Political Economy, 78*(20), 311–329.

Nitzl, C., Roldán, J. L., & Cepeda Carrión, G. (2016). Mediation analysis in partial least squares path modeling: Helping researchers discuss more sophisticated models. *Industrial Management & Data Systems, 116*(9), 1849–1864.

Office for National Statistics (UK). (n.d.). Online banking penetration in Great Britain from 2007 to 2017. In Statista—the statistics Portal. Retrieved March 15, 2017, from https://www.statista.com/statistics/286273/internet-banking-penetration-in-great-britain/.

Palmatier, R. W. (2008). *Relationship marketing*. Cambridge: Marketing Science Institute.

Palmatier, R. W., Dant, R. P., Grewal, D., & Evans, K. R. (2006). Factors influencing the effectiveness of relationship marketing: A meta-analysis. *Journal of Marketing, 70*(4), 136–153.

Podolny, J. M., & Phillips, D. J. (1996). The dynamics of organizational status. *Industrial and Corporate Change, 5*(2), 453–471.

Rindova, V. P., Williamson, I. O., Petkova, A. P., & Sever, J. M. (2005). Being good or being known: An empirical examination of the dimensions, antecedents, and consequences of organizational reputation. *Academy of Management Journal, 48*(6), 1033–1049.

Ringle, C. M., Wende, S., & Will, A. (2010). Finite mixture partial least squares analysis: Methodology and numerical examples. In V. E. Vinzi, W. W. Chin, J. Henseler, & H. Wang (Eds.), *Handbook of partial least squares* (pp. 195–218). Heidelberg: Springer.

Ringle, C. M., Wende, S., & Becker, J.-M. (2015). *SmartPLS 3*. Boenningstedt: SmartPLS GmbH. http://www.smartpls.com.

Romanian Banking Association. (2016). *The banking sector of Romania*. Retrieved March 20, 2017, from http://www.arb.ro/en/romanian-banking-system/romanian-banking-system-2/.

Rosenzweig, P. M. (1994). When can management science research be generalized internationally? *Management Science, 40*(1), 28–39.

Ruiz, B., García, J. A., & Revilla, A. J. (2016). Antecedents and consequences of bank reputation: A comparison of the United Kingdom and Spain. *International Marketing Review, 33*(6), 781–805.

Sarstedt, M. (2008). A review of recent approaches for capturing heterogeneity in partial least squares path modelling. *Journal of Modelling in Management, 3*(2), 140–161.

Sarstedt, M., & Ringle, C. M. (2010). Treating unobserved heterogeneity in PLS path modeling: A comparison of FIMIX-PLS with different data analysis strategies. *Journal of Applied Statistics, 37*(8), 1299–1318.

Sarstedt, M., & Schloderer, M. P. (2010). Developing a measurement approach for reputation of non-profit organizations. *International Journal of Nonprofit and Voluntary Sector Marketing, 15*(3), 276–299.

Sarstedt, M., Schwaiger, M., & Ringle, C. M. (2009). Determining the number of segments in FIMIX-PLS. In L. Robinson (Ed.), *Developments in marketing science: Proceedings of the academy of marketing science*. Baltimore: Springer.

Sarstedt, M., Becker, J.-M., Ringle, C. M., & Schwaiger, M. (2011). Uncovering and treating unobserved heterogeneity with FIMIXPLS: Which model selection criterion provides an appropriate number of segments? *Schmalenbach Business Review, 63*(1), 34–62.

Sarstedt, M., Ringle, C. M., Henseler, J., & Hair, J. F. (2014). On the emancipation of PLS-SEM: A commentary on Rigdon (2012). *Long Range Planning, 47*(3), 154–160.

Schlägel, C., & Sarstedt, M. (2016). Assessing the measurement invariance of the four-dimensional cultural intelligence scale across countries: A composite model approach. *European Management Journal, 34*(6), 633–649.

Schwaiger, M. (2004). Components and parameters of corporate reputation-an empirical study. *Schmalenbach Business Review, 56*(1), 46–71.

Shamma, H. M. (2012). Toward a comprehensive understanding of corporate reputation: Concept, measurement and implications. *International Journal of Business and Management, 7*(16), 151–169.

Shamma, H. M., & Hassan, S. S. (2009). Customer and non-customer perspectives for examining corporate reputation. *Journal of Product & Brand Management, 18*(5), 326–337.

Shiau, W. L., & Luo, M. M. (2012). Factors affecting online group buying intention and satisfaction: A social exchange theory perspective. *Computers in Human Behavior, 28*(6), 2431–2444.

Shore, L. M., Tetrick, L. E., Lynch, P., & Barksdale, K. (2006). Social and economic exchange: Construct development and validation. *Journal of Applied Social Psychology, 36*(4), 837–867.

Smith, J. B., & Barclay, D. W. (1997). The effects of organizational differences and trust on the effectiveness of selling partner relationships. *Journal of Marketing, 61*(1), 3–21.

Steenkamp, J. B. E., & Baumgartner, H. (1998). Assessing measurement invariance in cross-national consumer research. *Journal of Consumer Research, 25*(1), 78–90.

Stiglitz, J. E. (2002). Information and the change in the paradigm in economics. *American Economic Review, 92*(3), 460–501.

Terblanche, N. S. (2014). Validation of the customer-based corporate reputation scale in a retail context. *International Journal of Market Research, 56*(5), 655–671.

The Economist Intelligence Unit. (2015). *Democracy Index 2015: Democracy in an age of anxiety*. Retrieved March 15, 2017, from https://www.yabiladi.com/img/content/EIU-Democracy-Index-2015.pdf.

The UK Cards Association. (2015). *UK card payments*. Retrieved March 15, 2017, from http://www.theukcardsassociation.org.uk/wm_documents/UK%20Card%20Payments%202015%20taster%20for%20website.pdf.

The World Bank. (2013). *World development indicators*. Retrieved March 15, 2017, from http://databank.worldbank.org/data/.

The World Bank. (2015). *The worldwide governance indicators (WGI)*. Retrieved March 15, 2017, from http://info.worldbank.org/governance/wgi/index.aspx#reports.

Transparency International. (2015). *Corruption perceptions index 2015*. Retrieved March 15, 2017, from https://www.transparency.org/cpi2015/#results-table.

Vidaver-Cohen, D. (2007). Reputation beyond the rankings: A conceptual framework for business school research. *Corporate Reputation Review, 10*(4), 278–304.

Vinzi, V. E., Trinchera, L., & Amato, S. (2010). PLS path modeling: From foundations to recent developments and open issues for model assessment and improvement. In V. E. Vinzi, W. W. Chin, J. Henseler, & H. Wang (Eds.), *Handbook of partial least squares* (pp. 47–82). Heidelberg: Springer.

Walker, K. (2010). A systematic review of the corporate reputation literature: Definition, measurement, and theory. *Corporate Reputation Review, 12*(4), 357–387.

Walker, K., & Dyck, B. (2014). The primary importance of corporate social responsibility and ethicality in corporate reputation: An empirical study. *Business and Society Review, 119*(1), 147–174.

Walsh, G., & Beatty, S. E. (2007). Customer-based corporate reputation of a service firm: Scale development and validation. *Journal of the Academy of Marketing Science, 35*(1), 127–143.

Walsh, G., & Wiedmann, K. P. (2004). A conceptualization of corporate reputation in Germany: An evaluation and extension of the RQ. *Corporate Reputation Review, 6*(4), 304–312.

Walsh, G., Mitchell, V. W., Jackson, P. R., & Beatty, S. E. (2009a). Examining the antecedents and consequences of corporate reputation: A customer perspective. *British Journal of Management, 20*(2), 187–203.
Walsh, G., Beatty, S. E., & Shiu, E. M. (2009b). The customer-based corporate reputation scale: Replication and short form. *Journal of Business Research, 62*(10), 924–930.
Walsh, G., Bartikowski, B., & Beatty, S. E. (2014). Impact of customer-based corporate reputation on non-monetary and monetary outcomes: The roles of commitment and service context risk. *British Journal of Management, 25*(2), 166–185.
Wang, Y., Lo, H. P., & Hui, Y. V. (2003). The antecedents of service quality and product quality and their influences on bank reputation: Evidence from the banking industry in China. *Managing Service Quality: An International Journal, 13*(1), 72–83.
Wang, Y., Kandampully, J. A., Lo, H. P., & Shi, G. (2006). The roles of brand equity and corporate reputation in CRM: A Chinese study. *Corporate Reputation Review, 9*(3), 179–197.
Weigelt, K., & Camerer, C. (1988). Reputation and corporate strategy: A review of recent theory and applications. *Strategic Management Journal, 9*(5), 443–454.
Woodside, A. G. (2016). The good practices manifesto: Overcoming bad practices pervasive in current research in business. *Journal of Business Research, 69*(2), 365–381.
World Justice Project. (2015). *Rule of law index 2015*. Retrieved March 15, 2017, from https://worldjusticeproject.org/sites/default/files/roli_2015_0.pdf.
Zhang, Y. (2009). A study of corporate reputation's influence on customer loyalty based on PLS-SEM model. *International Business Research, 2*(3), 28–35.
Zhao, X., Lynch, J. G., & Chen, Q. (2010). Reconsidering Baron and Kenny: Myths and truths about mediation analysis. *Journal of Consumer Research, 37*(2), 197–206.

Chapter 5
The Compliance Index Model: Mitigating Compliance Risks by Applying PLS-SEM to Measure the Perceived Effectiveness of Compliance Programs

Sebastian Rick and Ralf Jasny

Abstract The Compliance Index Model represents a new type of employee-based measurement system for evaluating—and enhancing—the effectiveness of compliance programs. The authors discuss the nature and purpose of the Compliance Index Model and explain the theory underlying the model, and the quantitative approach employed to estimate the model parameters. The authors also illustrate the use of importance-performance analysis to identify critical areas of managerial attention and action for improving compliance program effectiveness and, consequently, mitigating compliance risk. Highlights of the findings include that (1) misaligned incentives, role-model-behavior, and transparency are factors that determine overall compliance culture, (2) overall compliance culture determines whistleblower readiness and compliance risk, and (3) compliance culture provides the foundation for compliance program effectiveness. The authors conclude with a discussion of the implications of the Compliance Index Model for public policy makers, managers, customers, as well as finance and banking in general.

Keywords Compliance Index Model · Compliance Programs · Performance Measurement · Effectiveness · Optimization

---

S. Rick (✉)
Governance and Assurance Services, KPMG AG Wirtschaftsprüfungsgesellschaft, The SQUAIRE, Frankfurt, Germany
e-mail: s.rick1983@gmail.com

R. Jasny
Faculty 3: Business and Law, Frankfurt University of Applied Sciences, Frankfurt, Germany
e-mail: rjasny@aol.com

N.K. Avkiran, C.M. Ringle (eds.), *Partial Least Squares Structural Equation Modeling*, International Series in Operations Research & Management Science 267, https://doi.org/10.1007/978-3-319-71691-6_5

5.1 Introduction

Despite establishment of clear and unambiguous behavioral norms to ensure regulatory compliance in firms, compliance risks in recent years have become increasingly a subject of concern for banks and other financial institutions. A recent study by the Boston Consulting Group (BCG) reported by Bloomberg revealed that, since the 2008 financial crisis, banks and other financial institutions globally have paid a total of 321 billion US dollars in penalties for violating laws, regulations, standards, etc. The report also held that the sum is likely to continue to mount in coming years. To date, especially the US authorities have levied high penalties, with the European supervisory authorities close behind. Just this past year, banks globally were fined 42 billion US dollars—a 68% increase over 2015. Per the BCG report, the sanctioned offenses range from violations of assorted laws to money laundering, financing terrorism and market manipulation. The BCG analysis further holds out scant hope that banks will see significant relief under the new US president. Donald Trump may have promised to relax or revoke numerous financial industry regulations, but, say the experts, by no means is the "era of steadily growing regulation over." Going forward, "penalties and the related legal costs" will continue to be a significant cost factor for the banking industry.

To ensure that laws, regulations, standards, etc. are followed and lapses are avoided, banks and other financial institutions have implemented internal compliance programs comprised of various practical measures and instruments, such as a code of conduct, a compliance officer, a training concept as well as monitoring and control measures. Compliance programs, variously also called ethics programs (Kaptein 2014), integrity programs (Blodgett 2011), ethical compliance programs (McKendall et al. 2002), business ethics and compliance programs (Joseph 2002), shared value programs (Weaver and Treviño 1999) or responsible behavior programs (Weaver and Treviño 1999), can be described generally as a formal control systems deployed in firms to improve the compliance culture, avoid lapses and foster rule-compliant behavior (Kedia et al. 2016; Kaptein 2014, 2009; McKendall et al. 2002; Weaver et al. 1999a, b). Hence, the compliance program's aim is to influence employee behavior, i.e., promote rule-compliant behavior and prevent infractions.

Despite the surge in compliance program implementations, however, there is little documented empirical evidence of their effectiveness. Thus, Kaptein (2014), for example, states that research to date, besides examining public entities such as universities, (Weber 2006), hospitals (Tobin et al. 2005) and government institutions (Beeri et al. 2013; Park and Blenkinsopp 2013), has concentrated mostly on private firms. But, in those cases, the investigation focuses either on just one firm or a small sample of firms (MacLean and Behnam 2010; Pelletier and Bligh 2006; Treviño and Weaver 2001; Weaver and Treviño 1999; Paine 1994; Barker 1993), only certain lapses, e.g., fraud (Smith et al. 2014), illegality (McKendall et al. 2002), conflicts of interest (Felo 2001) and inconsistent values (Majluf and Navarrete 2011), or else only on particular compliance program components like behavior/ethics codes (Kaptein 2011a; Kaptein and Schwartz 2008) and training

(Delaney and Sockell 1992; Warren et al. 2014). Other studies on the effectiveness of compliance programs instead focus exclusively on the business culture (Kaptein 2011b) as such or the employees' ethical intent (Ruiz et al. 2014). Moreover, studies on the effectiveness of compliance programs were conducted not on lapses per se, but rather, for example, on job satisfaction (Valentine and Fleischman 2008), on readiness of corporate social responsibility (Valentine and Fleischman 2008), and on external communication (Felo 2007). And, a small number of studies focused on the effectiveness of compliance program types from the perspective of whether a compliance program should be more rule- or value oriented (Weaver and Treviño 1999) or rather decoupled or integrated (MacLean et al. 2014).

However, not only do scienctists face problems to validate the effectiveness of compliance programs, but so do practioneers. Unlike many corporate functions, compliance officers cannot directly point to bottom-line savings, increased sales, or the like to demonstrate the effectiveness of the compliance program. To partially deal with this difficulty, over the years compliance officers have become more sophisticated in what they count and how they interpret the numbers. For example, they divide calls to the hotline into categories such as calls for advice vs. calls to report problems, and anonymous calls vs. calls that identify the caller. They consider increased calls for advice a good thing, and evidence that training is working. They also consider increases in calls that identify the caller (vs. anonymous calls) to be a sign that the caller trusts them to deal with the problem appropriately, without negative repercussions for the caller. On surveys, compliance officers look for improved perceptions of role-model behavior. Still, they felt that such attempts to measure and demonstrate compliance program effectiveness are limited in their value. For example, interpreting increases in reports of misconduct remains ambiguous—is that a good thing (in that people feel more comfortable reporting it) or a bad thing (in that it might reflect an increase in dubious behavior)? So, even after reporting on these metrics to executives and the board, ambiguity about the effectiveness of the compliance program remains high.

Our goal here, therefore, is to develop an empirically grounded quantitative model for banks and other financial institutions to measure under which conditions employees perceive compliance programs as being effective and what consequences can be extrapolated from this perception, particularly as concerns the employees' readiness to report rule-breaking behavior and the mitigation of compliance risks. In doing so, we advance the fundamental hypothesis that an organization's compliance culture provides the foundation for the compliance program's perceived effectiveness. To put it differently: The compliance culture influences the importance that employees of banks and other financial institutions attach to following the rules and thus how prepared they are to behave in a rule-conforming manner and/or how tolerant they are of rule infractions. Closing this gap in the research is of critical importance, given that current regulatory pressure will leave a lasting imprint on the global financial industry and thus calls for a change of perspective. Confronted by the steadily rising regulatory demands, banks and other financial institutions must find a professional objeticve, reliable, and valid approach for adapting to this new era. Only then will they remain economically viable for the long run.

5.2 The Compliance Index Model and Methodology

The concept behind the Compliance Index Model, namely, a measure of overall compliance culture that is uniform and comparable, requires a methodology with two fundamental properties. First, the methodology must recognize that compliance culture and the other constructs in the model represent different types of employee perceptions that cannot be measured directly. Accordingly, the Compliance Index Model uses a multiple indicator approach to measure overall compliance culture as a latent variable. The result is the Compliance Index, which is general enough to be comparable across different banks and other financial institutions.

Second, as an overall measure of compliance culture, the Compliance Index Model must be designed in a way that not only accounts for employee experience but that is also forward-looking. To this end, the Compliance Index is embedded in a system of cause and effect relationships, which makes it the centerpiece in a chain of relationships running from the antecedents of overall compliance culture—incentives, role-model behavior and transparency—to the consequences of overall compliance culture—whistleblower readiness and compliance risk. This design allows the Compliance Index to capture the served employee's perception of a firm's compliance culture in a manner that is both backward- and forward-looking. Moreover, modeling the Compliance Index as part of such a system serves to validate the Compliance Index from a nomological standpoint. Nomological validity, a form of construct validity, is the degree to which a construct behaves as predicted within a system of related constructs called a nomological net (Cronbach and Meehl 1955). To the extent that the Compliance Index Model predictions are supported, the validity of the Compliance Index is supported.

5.2.1 Model Specification

The Compliance Index Model is a complete structural equation model consisting of one structural model and two measurement models, each of which can also be differentiated into a *reflective* and a *formative* measurement model. Thus, the Compliance Index Model can be described as the general case of the structural equation analysis that contains a structural equation model with manifest variables as the special case (Weiber and Mühlhaus 2014, p. 36). All latent endogenous variables are identified in the structural model with the abbreviation η and all latent exogenous variables with ξ.

The variables notation of the Compliance Index Model is depicted in summary form in Fig. 5.1, also already showing the abbreviations for the measurement variables (manifest variables) and of the residuals.

In what follows, we first formulate the path diagrams of the three component models of the Compliance Index Model. Building on this, the path diagrams are next transformed into a multi-equation system, which becomes the basis for computing the actual model using the PLS approach or, specifically, the Smart PLS 3 software program developed by Ringle et al. (2015).

Abbreviation	Name	Meaning
η	Eta	Latent endogenous variable
ξ	Ksi	Latent exogenous variable
y	Ypsilon	Manifest measurement variable of a latent endogenous variable
x	Iks	Manifest measurement variable of a latent exogenous variable
ε	Epsilon	Residual of a reflective measurement variable$y$
ζ	Zeta	Residual of latent endogenous variable

Fig. 5.1 Notations for variable in the Compliance Index Model

5.2.2 *Specifying the Path Model*

The structural model represents the theoretically assumed relationships between the latent endogenous variables "role model behavior"(η_1), "transparency" (η_2), "compliance culture" (η_3), "whistleblower readiness" (η_4) and "compliance risk" (η_5) and the latent exogenous variable "incentives"(ξ_1). In the process, the latent endogenous variables are explained by the assumed causal relationships in the model, wherein the latent exogenous variable serves as the explanatory factor but is itself not explained by the structural model.

The predictive/causal relationships between the individual variables are represented in the structural model by one-way arrows. The path coefficients shown represent the direct causal effects between the variables and, in each case, indicate the strength and effect direction of this relationship. They are indicated and abbreviated, respectively, as β and γ.

Since the structural model's latent variables are immune to direct empirical observation, capturing their real-world characteristics requires operationalizing them with measurement models (Fig. 5.2).

5.2.3 *Specifying the Measurement Models*

The reflective and the formative measurement model both contain formal directions (operations) for how the structural model's individual latent variables are to be captured and measured with numbers (measurement instruction).[1] In the process, each direction of causality (reflective vs. formative) is specified by the formulation and choice of measurement indicators. It follows that the reflective measurement model

[1] See Appendix 5.1 to this chapter for the operationalization of the structural model's latent variables using the measured indicators.

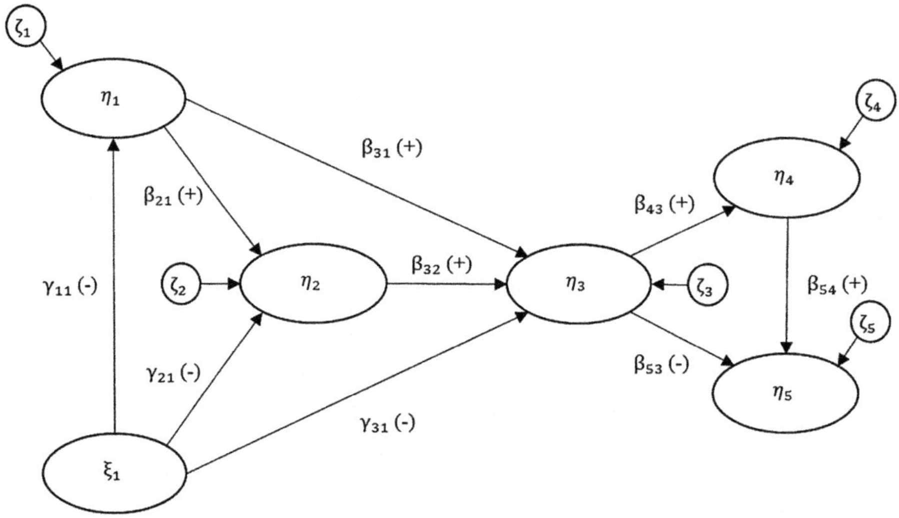

Fig. 5.2 The structural model path diagram

contains the empirical measurement values from operationalizing the latent endogenous variables and mirrors the assumed relationships between the measured values and the endogenous items. By contrast, the formative measurement model contains the empirical measured values yielded by operationalizing the exogenous variables and reflects the assumed relationships between the measured values and exogenous item.

In the reflective measurement model, the latent endogenous variables represent the cause of the reflective measurement indicators or measurement variables $y' = (y_1, y_2, \ldots, y_p)$ to be captured on the observational level. Accordingly, the reflective measurement indicators mirror observable outcomes or consequences of the effectiveness of the latent endogenous variables. By contrast, in the formative measurement model the latent exogenous variables are understood as the consequence of the formative measurement indicators or measurement variables $x' = (x_1, x_2, \ldots, x_q)$ operative on the observational level. The latent exogenous variable therefore represents an exact (measurement-error free) linear combination of the formative measurement indicators (e.g., Diamantopoulos 2011; Edwards and Bagozzi 2000; Fornell and Bookstein 1982).

As shown in Figs. 5.3 and 5.4, each direction of causality of the measurement models is represented by a one-way arrow to indicate a predictive/causal relationship between the individual variables. The pictured factor loads and multiple regression coefficients here represent the outer weighting factors and are abbreviated by the notation λ (outer loadings) in the reflective measurement model and by π (outer weights) in the formative measurement model.

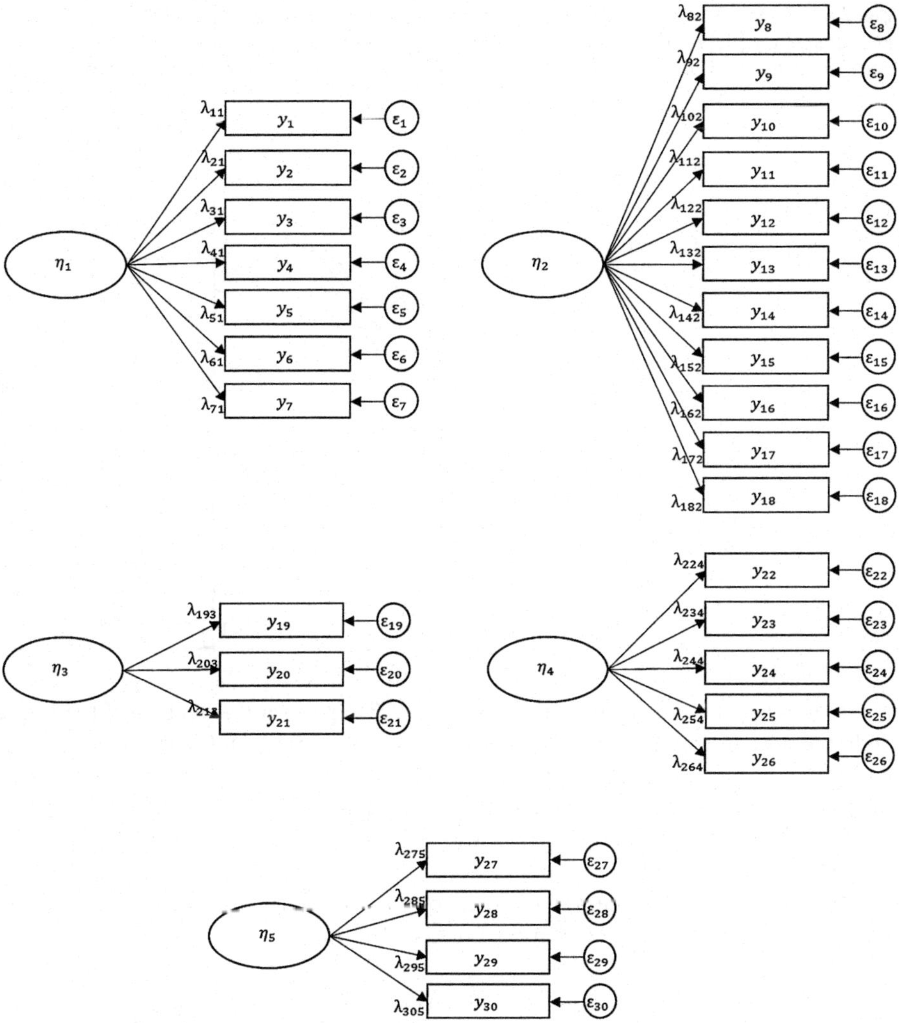

Fig. 5.3 Reflective measurement model of the latent endogenous variables

5.2.4 The Compliance Index Model Equations

After Fornell et al. (1996), we can describe the theoretically assumed relationships between the latent variables in the structure model mathematically as a formal series of equations that will be estimated using the PLS approach, so that the systematic part of the predictor relationships corresponds to the conditional expected value of the latent endogenous variables. With that, the general equation of the structural model is stochastically specified and now reads as follows:

$$E[\eta | \eta\xi] = B\eta + \Gamma\xi$$

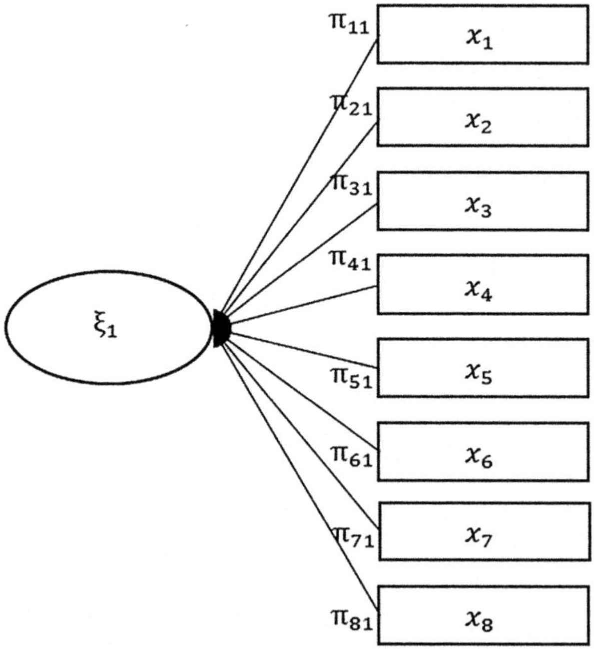

Fig. 5.4 Formative measurement model of the latent exogenous variables

where $\eta' = (\eta_1, \eta_2, \ldots, \eta_m)$ and $\xi x' = (\xi_1, \xi_2, \ldots, \xi_n)$represent the vectors of the latent endogenous and exogenous variables and B $(m\ x\ m)$ correspond to the matrix of the partial regression coefficients (path coefficients) for η and also Γ $(m\ x\ n)$ the matrix of the partial regression coefficients (path coefficients) for ξ. From Fornell et al. (1996, p. 14) it follows that $E[\eta\zeta'] = E[\xi\zeta'] = E[\zeta] = 0$ with $\zeta = \eta - E[\eta|\ \eta\xi]$ as vector of the residuals.

Thus, the specific equations that describe the effect relationships between the latent variables in the structural model read as follows:

$$\begin{pmatrix} \eta_1 \\ \eta_2 \\ \eta_3 \\ \eta_4 \\ \eta_5 \end{pmatrix} = \begin{pmatrix} 0 & 0 & 0 & 0 & 0 \\ \beta_{21} & 0 & 0 & 0 & 0 \\ \beta_{31} & \beta_{32} & 0 & 0 & 0 \\ 0 & 0 & \beta_{43} & 0 & 0 \\ 0 & 0 & \beta_{53} & \beta_{54} & 0 \end{pmatrix} \begin{pmatrix} \eta_1 \\ \eta_2 \\ \eta_3 \\ \eta_4 \\ \eta_5 \end{pmatrix} + \begin{pmatrix} \gamma_{11}\alpha_1 \\ \gamma_{21}\alpha_2 \\ \gamma_{31}\alpha_3 \\ 0\ \alpha_4 \\ 0\ \alpha_5 \end{pmatrix} \xi_1 + \begin{pmatrix} \zeta_1 \\ \zeta_2 \\ \zeta_3 \\ \zeta_4 \\ \zeta_5 \end{pmatrix}.$$

After Fornell et al. (1996, p. 16) and Diamantopoulos (2011, p. 337), the general equations that describe the effect relationships between the latent variables in the structural model and the directly observable measured variables in the measurement models read as

$$y = \Lambda_y\eta + \varepsilon$$

and

$$\xi = \Pi_x x$$

where $y' = (y_1, y_2, \ldots, y_p)$ and $x' = (x_1, x_2, \ldots, x_q)$ correspond to the directly observable measured variables and $\Lambda_y(p\ x\ m)$ and $\Pi_x(q \times n)$ to the regression matrices. From Fornell et al. (1996, p. 14f.; 1982, p. 443) it follows that $E[\varepsilon] = E[y\varepsilon'] = 0$.

Therefore, the specific equations that describe the effect relationships between the latent variables and the directly observable measured variables in the measurement models read as follows:

$$\begin{pmatrix} y_1 \\ y_2 \\ y_3 \\ y_4 \\ y_5 \\ y_6 \\ y_7 \\ y_8 \\ y_9 \\ y_{10} \\ y_{11} \\ y_{12} \\ y_{13} \\ y_{14} \\ y_{15} \\ y_{16} \\ y_{17} \\ y_{18} \\ y_{19} \\ y_{20} \\ y_{21} \\ y_{22} \\ y_{23} \\ y_{24} \\ y_{25} \\ y_{26} \\ y_{27} \\ y_{28} \\ y_{29} \\ y_{30} \end{pmatrix} = \begin{pmatrix} \lambda_{11} & 0 & 0 & 0 & 0 \\ \lambda_{21} & 0 & 0 & 0 & 0 \\ \lambda_{31} & 0 & 0 & 0 & 0 \\ \lambda_{41} & 0 & 0 & 0 & 0 \\ \lambda_{51} & 0 & 0 & 0 & 0 \\ \lambda_{61} & 0 & 0 & 0 & 0 \\ \lambda_{71} & 0 & 0 & 0 & 0 \\ 0 & \lambda_{81} & 0 & 0 & 0 \\ 0 & \lambda_{92} & 0 & 0 & 0 \\ 0 & \lambda_{102} & 0 & 0 & 0 \\ 0 & \lambda_{112} & 0 & 0 & 0 \\ 0 & \lambda_{122} & 0 & 0 & 0 \\ 0 & \lambda_{132} & 0 & 0 & 0 \\ 0 & \lambda_{142} & 0 & 0 & 0 \\ 0 & \lambda_{152} & 0 & 0 & 0 \\ 0 & \lambda_{162} & 0 & 0 & 0 \\ 0 & \lambda_{172} & 0 & 0 & 0 \\ 0 & \lambda_{182} & 0 & 0 & 0 \\ 0 & 0 & \lambda_{193} & 0 & 0 \\ 0 & 0 & \lambda_{203} & 0 & 0 \\ 0 & 0 & \lambda_{213} & 0 & 0 \\ 0 & 0 & 0 & \lambda_{224} & 0 \\ 0 & 0 & 0 & \lambda_{234} & 0 \\ 0 & 0 & 0 & \lambda_{244} & 0 \\ 0 & 0 & 0 & \lambda_{254} & 0 \\ 0 & 0 & 0 & \lambda_{264} & 0 \\ 0 & 0 & 0 & 0 & \lambda_{275} \\ 0 & 0 & 0 & 0 & \lambda_{285} \\ 0 & 0 & 0 & 0 & \lambda_{295} \\ 0 & 0 & 0 & 0 & \lambda_{305} \end{pmatrix} \begin{pmatrix} \eta_1 \\ \eta_2 \\ \eta_3 \\ \eta_4 \\ \eta_5 \end{pmatrix} + \begin{pmatrix} \varepsilon_1 \\ \varepsilon_2 \\ \varepsilon_3 \\ \varepsilon_4 \\ \varepsilon_5 \\ \varepsilon_6 \\ \varepsilon_7 \\ \varepsilon_8 \\ \varepsilon_9 \\ \varepsilon_{10} \\ \varepsilon_{11} \\ \varepsilon_{12} \\ \varepsilon_{13} \\ \varepsilon_{14} \\ \varepsilon_{15} \\ \varepsilon_{16} \\ \varepsilon_{17} \\ \varepsilon_{18} \\ \varepsilon_{19} \\ \varepsilon_{20} \\ \varepsilon_{21} \\ \varepsilon_{22} \\ \varepsilon_{23} \\ \varepsilon_{24} \\ \varepsilon_{25} \\ \varepsilon_{26} \\ \varepsilon_{27} \\ \varepsilon_{28} \\ \varepsilon_{29} \\ \varepsilon_{30} \end{pmatrix}$$

and

$$\xi_1 = \left(\pi_{11}\ \pi_{21}\ \pi_{31}\ \pi_{41}\ \pi_{51}\ \pi_{61}\ \pi_{71}\ \pi_{81}\right)\begin{pmatrix} x_1 \\ x_2 \\ x_3 \\ x_4 \\ x_5 \\ x_6 \\ x_7 \\ x_8 \end{pmatrix}.$$

It follows from Fornell and Bookstein (1982, p. 443) that $E[\zeta\zeta'] = \Psi$ and $E[\varepsilon\varepsilon'] = \Theta_\varepsilon$, so that the PLS algorithm can minimize the sum of the diagonal values (trace) of Ψ and Θ_ε. Since the model's elements that do not lie on the diagonal are known and the constructed values of the latent variables are explicitly estimated in the PLS approach, no identification problems exist for recursively formulated PLS models. Moreover, because no assumptions need to be made for PLS estimates about the population and the measurement domain estimates, there are also no demands on the dispersion.

Finally, it is specified that the model variables will be z-standardized, so that $E[\eta] = E[\xi] = 0$ and $Var(\eta) = Var(\xi) = Var(y) = Var(x) = 1$ and $E[y] = E[x] = 0$ applies. By using path analysis, it is possible to estimate the effect magnitudes in the structural model (B and Γ) based on the empirical correlation or covariance matrix of the latent variables.[2]

5.2.5 Empirical Study

Between June and August 2016, using a standardized anonymous questionnaire, we surveyed in writing a total of 200 employees in 40 different German banks and other financial institutions to study the assumed relationships between the latent variables of the structural model empirically.[3] The employees were not compensated for participating. The chosen survey method, on the one hand, ensured that there was no chance of the interviewer influencing the survey subject's answers, thus preventing any interview effects (Biemer and Lyberg 2003, p. 171). On the other hand, it also served to make it impossible to trace the survey subjects' answers back to them. For the employee survey, we used an improved version of an empirically validated questionnaire developed by Kaptein (2008). The employees used stamped, pre-addressed

[2] Due to z-standardization of the initial variables, the empirical covariance matrix and the correlation matrix of the latent variables are identical.

[3] The actual administering of the employee survey was oriented in part on the recommendations for written mailed surveys made by GESIS—Leibniz Institute for the Social Sciences (Menold 2014).

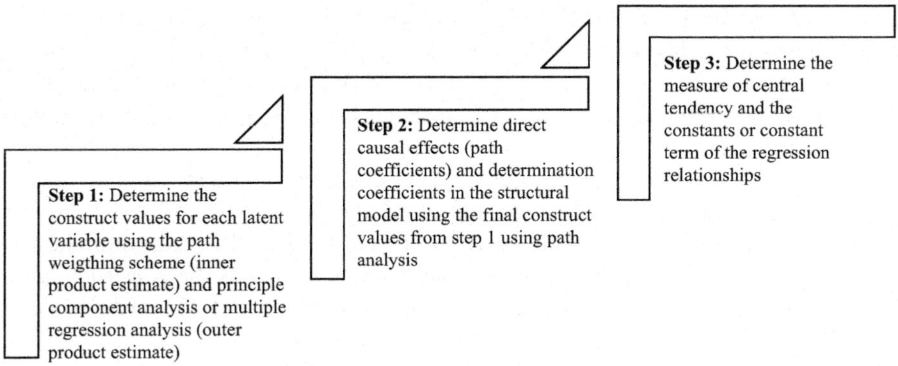

Fig. 5.5 Consecutive steps in PLS estimation

return envelopes furnished by us to return the completed questionnaires to us. The response rate was 42.5%, equal to 85 filled-out questionnaires. The response quotient was 2.5% higher than is usual for a written, mailed survey, per a meta-analysis by Shih and Fan (2008).

5.2.6 Model Estimation and the PLS-SEM Algorithm

The model estimation conforms to the formal model specification based on the empirical study with the goal of achieving an optimum prediction of the data matrix with reference to the endogenous construct "compliance risk". To this end, the parameters of the Compliance Index Model were estimated with the help of the PLS approach to minimize the discrepancy between empirical and estimated case data. Per Dijkstra (2010, p. 3), the PLS algorithm employed in the process can be conceived of as a series of independent *ordinary least squares* (OLS) regressions in the form of weighted vectors that, after their convergence, satisfy fixed point equations.

In specific terms, estimating the Compliance Index Model parameters with the SmartPLS 3 software developed by Ringle et al. (2015), is done in three consecutive steps that are summarized in Fig. 5.5.[4] During the first stage, using the initial data, actual construct values are estimated for each latent variable in every captured case (each surveyed individual) using the path weighting scheme (inner product estimate) and principal component analysis or multiple regression analysis (outer product estimate) by means of an iterative process of two-way fitting of the data from the measurement models (Λ_y and $\prod_x$) and the structural model (B and Γ). Once the construct values of the latent variables are finalized, they are then used in the second

[4]The description that follows of the three consecutive steps in model estimation is based on Hair et al. (2016, p. 82–92), who offer a relevant, practical introduction to SmartPLS 3 by Ringle et al. (2015).

stage to estimate the path coefficients in the structural model and the determination coefficients of the latent variables using path analysis. When the effect sizes and determination coefficients are also final, the third stage concludes with determining the measure of central tendency and the constants or the constant term in the regression relationships.

To be able to estimate specific construct values for every latent variable in each surveyed case during the first stage, the initial data must first be loaded into the software program. It is incumbent on the user to instruct the PLS algorithm how to handle data missing from the initial data set when it estimates the construct value. The user basically can choose from three missing data settings: (1) mean substitution, (2) listwise deletion and (3) pair-wise deletion.[5]

- In the present actual use case, mean substitution has been selected for handling missing data as the algorithm as well as parameter settings in the software program.

Once the handling of missing data is set, the user must specify a convergence criterion and the maximum number of iterations. The convergence criterion ensures that the PLS algorithm stops as soon as the sum of the variances of all external weights between two consecutive iterations is smaller than the previously specified convergence criterion. Per Hair et al. (2016, p. 91), this convergence criterion should be sufficiently small and 10^{-7} (0.0000001) is recommended. The maximum number of iterations puts a limit on how often the inner and outer construct estimation is to be repeated in calculating the PLS results. After Hair et al. (2016, p. 91), this number should be sufficiently large to ensure the PLS algorithm converges and the recommended setting is 300 iterations.

- In the present actual use case, a convergence criterion of 10^{-7} (0.0000001) and a maximum of 300 iterations have been selected for the algorithm as well as parameter settings in the software program.

After setting these basic algorithm and parameter criteria, next the user must specify how to estimate the inner and outer construct values. For the inner construct value estimate, the user has three options: (1) centroid weighting scheme, (2) factor weighting scheme and (3) path weighting scheme. Hair et al. (2016, p. 89) here recommend the path weighting scheme for standard use, because it yields the highest value for the coefficient of determination R^2 for the latent variables and can be used generally for all types of PLS model specifications and estimates.[6]

- In the present actual use case, the path weighting scheme is selected as the algorithm and parameter settings inn the software program.

[5] For a general introduction to the algorithms described here for handling missing data, see, for example, Barladi and Enders (2010), Allison (2001), and Little and Rubin (1987) who offer a relevant overview.

[6] However, Johansson and Yip (1994, p. 587) note in this connection that the results that can be achieved with the three differing inner weighting schemes only diverge marginally from each other.

For estimating outer construct value, the user basically has two options: Mode A for reflective measurement models or Mode B for formative measurement models. The outer weigthing factors are determined *automatically* in the reflective measurement model through a somewhat special case of principal component analysis (PCA), wherein the found outer loadings (Λ_y) of the measurement variables are used as the outer weigthing factors (e.g., Weiber and Mühlhaus 2014, p. 71). In contrast, in the formative measurement model, the outer weighting factros are determined *automatically* through multiple regression analysis, wherein the found outer weights (Π_x) of the measurement variables are used as the outer weigthing factors. Hence, in Mode A, a simple regression model exists that has one latent variable as the independent and one measurement variable as the dependent variable, while in Mode B exists a multiple regression model with a latent variable as the dependent and a measurement variable as the independent variable. Here, the construct values obtained earlier from the inner estimate are used as values for the latent variables.

In contrast to the inner construct value estimation, for the outer construct value estimation the user must first set an initial starting value (starting approximation) for the outer weights. Hair et al. (2016, p. 90) state that +1 is a good value here in practice, which is why this value is also the default setting in the software program.[7] Once the outer weights for the latent variables have been found, based on information from the structural model, these are then used for calculating the outer estimate values of the latent variable. The inner and outer construct value estimation repeats until the convergence criterion is fulfilled (or the maximum number of iterations has been reached).

- In the present use case, +1 is selected for the algorithm and parameter settings in the software program as the initial starting value (starting approximation) for the outer weights as part of the outer construct value estimation.

Once the convergence criterion has been met (or the maximum number of iterations reached), the final construct values of the latent variables are found during the first step of the PLS algorithm. In the second step, these are then used to estimate the direct causal effects on the structural model and the determination coefficients of the latent variable with the help of path analysis based on information from the empirical covariance or correlation matrix of the latent variables. Per Wright (1921, 1923, 1934), the correlations between a dependent (criterion variable) and an independent variable (predictive variable) using path analysis can be segmented additively into a direct causal effect, an indirect causal effect, and a non-causal (correlative) effect. Here, the direct causal effects correspond to the path coefficients and the indirect causal effects result from multiplying the corresponding direct causal effects along the paths. The total causal effect then results from adding

[7] Per Hair et al. (2016, p. 90f.), Lohmöller recommends using +1 for all initial outer weights, except for the last measured variable of a measurement model, for which—1 is to be used. Reportedly this normally lets the PLS algorithm converge more rapidly. However, the so-called Lohmöller setting can produce unexpected sign changes in the estimated effect relationships in the measurement models and/or the structural model.

Setting	Selection
Data file settings	
Data file	Variables_10082016_PLS-MODEL [85 data sets]
Identifier for missing data	None
Data set-up settings	
Algorithm for handling missing data	Replace with mean
Weighting vector	-
PLS algorithm settings	
Metric data	Mean 0, variance 1
Weighting scheme	Path weighting scheme
Initial weights	1,0
Use Lohmoeller settings?	No
Max. number of iterations	300
Stop criterion	7
Outer weighting mode of constructs	
Incentives (ξ_1)	Automatic (Mode B; formative)
Role model behavior (η_1)	Automatic (Mode A; reflective)
Transparency (η_2)	Automatic (Mode A; reflective)
Compliance culture (η_3)	Automatic (Mode A; reflective)
Whistleblower readiness (η_4)	Automatic (Mode A; reflective)
Compliance risk (η_5)	Automatic (Mode A; reflective)

Fig. 5.6 PLS-algorithm and parameter settings in SmartPLS 3

the direct and indirect causal effect and can be immediately determined through the direct causal effects. Per Weiber and Mühlhaus (2014, p. 31), this segmentation is universally valid and independent of the structural model's size.

Finally, in the third step only the measure of central tendency and constants or the constant term $\alpha' = (\alpha_1, \alpha_2, \ldots, \alpha_n)$ of the regression relationships are calculated. However, since the model variables were z-standardized in the present actual use case, the path coefficients are to be interpreted as standardized partial regression coefficients and the constants or the constant term $\alpha' = (\alpha_1, \alpha_2, \ldots, \alpha_n)$ can therefore be omitted.

In conclusion, Fig. 5.6 summarizes once more the algorithm and parameter settings of the software program for running the model estimation. Once the PLS algorithm has run through all three steps without problems, the software program will automatically output a report on the resultant model estimation, which we will now discuss in the next part.[8]

5.3 Results

The results report generated by the software program essentially contains the result of the estimation of the path coefficients (direct causal effects) in the structural model (B and Γ) and the outer weighting factors (Λ_y and $\prod_x$) in the

[8] During model estimation, inconsistencies detected in the initial data or implausible parameter estimates are automatically recognized by technical check routines in the software program and result in aborting the parameter estimation with an error code output.

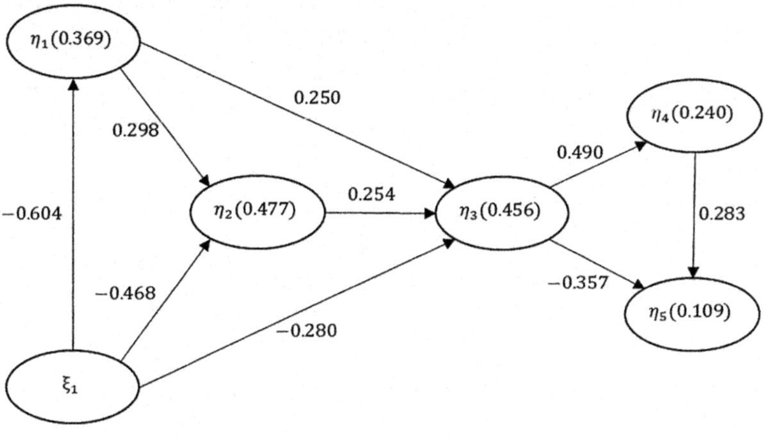

Fig. 5.7 PLS estimation of direct causal effects in the structural model

measurement models.[9] A check of the convergence information showed that the PLS algorithm was not simply terminated by reason of reaching the maximum number of possible iterations—the PLS algorithm converged after just 9 of 300 maximum allowed iterations.

5.3.1 Estimation Results

The results of estimating the path coefficients in the structural model (B and Γ) are summarized in Fig. 5.7, where the values depicted for the constructs "role model behavior" (η_1), "transparency" (η_2), "compliance culture" (η_3), "whistleblower readiness" (η_4) and "compliance risk" (η_5) correspond to the calculated determination coefficients. Illustration and description of the residuals (ζ) have been dispensed with here.

Comparing the estimated path coefficients with those previously imputed to those in the structural model, we see that all correspond to the assumed intensities and effect directions: so, for instance, an improvement in the compliance culture leads to a mitigation of the compliance risk (−0.357) and to an increased readiness by employees to report rule violations (0.490). The coefficients of determination, on the other hand, show that the three constructs "incentives," "role model behavior," and "transparency" explain 45.6% of the variance in construct "compliance culture". The rest of the path coefficients are summarized in Fig. 5.8.

[9] The following description of the model estimation results is limited to the estimated path coefficients (direct causal effects) in the structural model (B and Γ) and the outer weigthing factors (Λ_y and Π_x) in the measurement models. A description of the remaining results, such as that from the estimation of the covariance matrices of the residuals (Ψ und Θ_ε) is omitted from what follows.

Latent variables					
	Role model behavior (η_1)	Transparency (η_2)	Compliance culture (η_3)	Whistleblower readiness (η_4)	Compliance risk (η_5)
Incentives (ξ_1)	-.607	-.468	-.280		
Role model behavior (η_1)		.298	.250		
Transparency (η_2)			.254		
Compliance culture (η_3)				.490	-.357
Whistleblower readiness (η_4)					.283
Compliance risk (η_5)					

Fig. 5.8 PLS-estimate of path coefficients in the structural model

Direct causal effects exist whenever a construct directly influences another construct, as can be seen from the paths (causal arrows) between the constructs in Fig. 5.7. In contrast, indirect causal effects develop between constructs by a latent variable acting on another one through one or several intervening variables. While the intensity and direction of the direct causal effects can simply be read off the path coefficients entered next to the causal arrows, determining the indirect causal effects requires multiplying the relevant path coefficients: so, for example, an indirect causal effect exists between the compliance culture (η_3) and compliance risk (η_5), that is transmitted through whistleblower readiness (η_4). This indirect causal effect is calculated as follows:

$$\textit{Indirect effect}\ (\eta_3, \eta_5) = 0.490 \times 0.283 = 0.138$$

This means that the influence of compliance culture on compliance risk is also partially explained by the employees' readiness to report rule violations. "Whistleblower readiness" thus seems to represent a mechanism that reveals an additional, reverse relationship between "compliance culture" and "compliance risk." The remainder of the indirect causal effects are summarized in Fig. 5.9.

The direct and indirect causal effects then combine to yield the total causal effects between the latent variables. They are calculated as follows:

$$\textit{Total effect} = \textit{direct effect} + \textit{indirect effect}$$

For example, the total causal effect of compliance culture (η_3) on compliance risk (η_5) is calculated by adding the direct causal effect (−0.357) to the indirect causal effect (0.138) as follows:

Latent variables					
	Role model behavior (η_1)	Transparency (η_2)	Compliance culture (η_3)	Whistleblower readiness (η_4)	Compliance risk (η_5)
Incentives (ξ_1)		-.181	-.316	-.292	.131
Role model behavior (η_1)			.076	.159	-.071
Transparency (η_2)				.124	-.056
Compliance culture (η_3)					.138
Whistleblower readiness (η_4)					
Compliance risk (η_5)					

Fig. 5.9 PLS estimation of indirect causal effects in the structural model

$$\textit{Total effect}(\eta_3, \eta_5) = -0.357 + 0.138 = -0.219$$

Substantively, this result is interpreted to mean that, with every improvement in the compliance culture by an empirical unit, compliance risk drops by −0.219 scale points. Overall, the compliance culture therefore influences compliance risk negatively, that is, by the amount of the total causal effect of −0.219. The remaining causal effects are summarized in Fig. 5.10.

Figure 5.11 summarizes the results of estimating the outer weighting factors (Λ_y and Π_x) in the measurement models. It shows that all the outer loadings found are positive and have values ranging from 0.663 (y_9) to 0.932 (y_{27}) that are almost without exception >0.7. Hence, a change expressed in the construct "compliance culture" leads to a strong alteration in how the measurement variables y_{19} (0.829), y_{20} (0.920), y_{21} (0.869) manifest. Regarding the found outer weights, it emerges that their influence on the construct "incentives" varies in strength. For instance, a change in the expression of the x_2 measurement variable results in a strong modification (0.829) in the expression of the construct "incentives". In comparison, a change in the expression of the x_8 measurement variable leads to a relatively weak change in the character of the construct "incentives" (−0.063).

In conclusion, we can state that all estimated path coefficients in the structural model (B and Γ) as well as the found outer weighting factors (Λy and Π_x) in the measurement models are plausible and consistent with the theoretically and/or logically established causal relationships.

	Latent variables				
	Role model behavior (η_1)	Transparency (η_2)	Compliance culture (η_3)	Whistleblower readiness (η_4)	Compliance risk (η_5)
Incentives (ξ_1)	-.607	-.649	-.596	-.292	.131
Role model behavior (η_1)		.298	.326	.159	-.071
Transparency (η_2)			.254	.124	-.056
Compliance culture (η_3)				.490	-.219
Whistleblower readiness (η_4)					.283
Compliance risk (η_5)					

Fig. 5.10 PLS estimation of the total causal effects in the structural model

5.3.2 *Evaluating the Estimation Results*

The key goal of the causal analysis is an evaluation of the model estimation, since this is where we check if the relationships between the latent variables, *a priori* formulated based on theoretical and/or logical considerations, can be empirically confirmed by the survey data (e.g., Hair et al. 2016). Since the PLS approach is not equipped with a meaningful global criterion to apply in evaluating the model goodness of fit, it is recommended to take account of all the available individual criteria for evaluating the measurement models and the structural model in a kind of synopsis (Ringle 2004, p. 23). After Hair et al. (2016, p. 106), in doing so, the evaluation of the measurement models (outer goodness of fit test) and of the structural model (inner goodness of fit test) should be fundamentally differentiated as shown in Fig. 5.12, with the added recommendation to evaluate the measurement models before the structural model. This is an intrinsically logical procedure, since the goodness of fit of the parameter estimation of a structural model is materially determined by the goodness of fit of the measurement models. On the "garbage in—garbage out" principle, incorrectly measured constructs also lead to mistakes in estimating the relationships in the structural model. Consequently, the testing of the outer goodness of fit aims to assess the reliability and validity of the construct measurements, while the test of the inner goodness of fit seeks to appraise the path coefficients in the structural model as well as its explanatory and predictive power.

In accord with this two-step process, in what follows we will illustrate and describe, first, the result of testing the outer weigthing factors of the measurement models based on the outer loadings(Λ_y) and outer weights ($\prod_x$) and, second, the result from testing the path coefficients in the structural model (B and Γ).

Measured variable	Latent variables					
	Incentives (ξ_1)	Role model behavior (η_1)	Transparency (η_2)	Compliance culture (η_3)	Whistleblower readiness (η_4)	Compliance risk (η_5)
x_1	-.130					
x_2	.829					
x_3	.520					
x_4	-.121					
x_5	.122					
x_6	-.014					
x_7	-.226					
x_8	-.063					
y_1		.813				
y_2		.818				
y_3		.868				
y_4		.868				
y_5		.827				
y_6		.814				
y_7		.877				
y_8			.763			
y_9			.663			
y_{10}			.702			
y_{11}			.684			
y_{12}			.864			
y_{13}			.824			
y_{14}			.721			
y_{15}			.754			
y_{16}			.743			
y_{17}			.806			
y_{18}			.826			
y_{19}				.829		
y_{20}				.920		
y_{21}				.869		
y_{22}					.794	
y_{23}					.777	
y_{24}					.845	
y_{25}					.875	
y_{26}					.819	
y_{27}						.932
y_{28}						.923
y_{29}						.904
y_{30}						.858

Fig. 5.11 PLS estimates of the outer weigthing factors

5.3.2.1 Evaluation of the Measurement Models

To evaluate the outer weigthing factors (Λ_y and $\prod_x$) in what follows we will adduce the result from testing the outer weigthing factors (Λ_y and $\prod_x$) with the evaluation criteria depicted in Fig. 5.12. Here it is necessary to differentiate between reflectively and formatively measured constructs. The two approaches are based on different concepts and therefore also require taking into account differing evaluation criteria.[10]

[10] For an introduction to the goodness of fit criteria described in what follows for evaluating reflective and formative measurement models, see, for example, Hair et al. (2016, p. 106–122, 137–159).

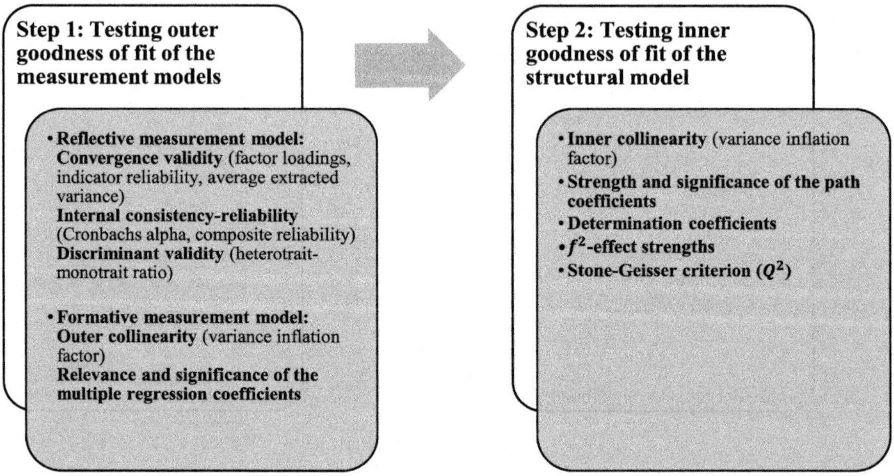

Fig. 5.12 Systematic evaluation of PLS estimation results

While the evaluation of reflective measurement models normally uses convergence validity (factor loadings, indicator reliability, average extracted variance), internal consistency reliability (Cronbach's alpha and composite reliability), and discriminant validity (heterotrait-monotrait ratio[11]), evaluating formative measurement models with these criteria is, as a rule, *not* possible, since they focus in particular on the correlations between the measurement variables, and formative measurement variables should be correlated as little as possible because of the collinearity problem (Hair et al. 2016, p. 138f.). Moreover, it is assumed with formative measurement models that they measure a construct precisely (free of measurement errors) (Diamantopoulos 2011, p. 337). For these reasons, usually other evaluation criteria are employed in evaluating formative measurement models that focus in particular on the collinearity between the measurement variables (variance inflation factor) and the relevance and significance of the found outer weights.[12] When doing so, non-significant regression coefficients should not be evaluated automatically as proof of scant measurement quality. Instead, Hair et al. (2016, p. 150) additionally recommend including in the evaluation the absolute contribution of a formative measurement variable to a construct (factor loading). These are

[11] According to Henseler et al. (2009), in the context of simulation studies performed, when it comes to evaluating the discriminant validity of PLS estimations, the typically used Fornell-Lacker criterion (Fornell and Lacker 1981) as well as various cross loading have proved to be unreliable. For this reason, both of these goodness of fit criteria have not been resorted to here.

[12] Normally, formatively measured constructs are also checked for content validity. It prevails when the measured indicators used represent the construct's contentual-sematic field and map to all defined significance contents of the construct (for example, Weiber and Mühlhaus 2014, p. 157). Typically, the proof of content validity is rendered by a careful choice of the individual measured indicators, expert opinions and/or pretests (Cronbach and Meehl 1955, p. 282; Nunnally 1967, p. 79 ff.).

conveniently routinely output by SmartPLS 3 by Ringle et al. (2015) regardless of the formative measurement model's specification orientation (Hair et al. 2016, p. 92). Only if the found outer weight and the outer loading are minimal or non-significant should eliminating a formative measurement variable be contemplated. However, since eliminating a measurement variable in formatively measure constructs always also results in them being measured only *incompletely*, doing so should never be based solely on static criteria—there should also always exist valid theoretical reasons for it (Henseler et al. 2009, p. 302; Weiber and Mühlhaus 2014, p. 265).

In the SmartPLS 3 software by Ringle et al. (2015), the evaluation criteria described are automatically output with the result report for every PLS model estimation or can be immediately and conveniently calculated with the help of the software program based on the model estimation results, thus obviating having to recalculate them with IBM SPSS or Microsoft Excel. Figures 5.13 and 5.14 summarize the evaluation of the measurement models.[13]

Thus, to begin with it, can be stated that the reflective measurement model overall manifests a very high convergence validity. Nearly all the found outer loadings with values ranging from 0.663 (y_9) to 0.932 (y_{27}) are above the recommended limit of ≥0.70. Solely the values of y_9 (0.663) and y_{11} (0.684) lie just below it. As expected, the indicator reliability values and the AEV are also high. In terms of the values of composite reliability and Cronbach's alpha for evaluating the internal consistency of the construct measurements, it also shows that overall the reflective model exhibits a very high internal consistency reliability. Both the values for composite reality and for Cronbach's alpha all clearly lie above the recommended limit of ≥0.60. Further, if we adduce the result of a two-sided significance test of the heterotrait-monotrait (HTMT) ratio of the correlations for evaluating the discriminant validity of the construct measurements in the reflective measurement model, it emerges that the HTMT ratio of the correlations diverges significantly from 1.[14] As a result, the constructs are "selectively" measured by the reflective measurement indicators and we have discriminant validity. If all recommended evaluation criteria are brought in, we can finally state that there is no need to eliminate a measurement variable. The outer loadings of both y_9 (0.663) and y_{11} (0.684) may lie just under the recommended limit of ≥0.70, but eliminating these two measurement variables has no appreciable effect on the internal consistency reliability of the construct "transparency", so that here we forego eliminating these two measurement variables (Hair et al. 2016, p. 114).[15] Hence, it can be declared for the reflective measurement model

[13] The referenced threshold values represent widely accepted standard values that are normally used in the literature. For more detailed information, see, for example, Hair et al. (2016, p. 106–122, 137–159) and/or Weiber and Mühlhaus (2014, p. 142, 155, 266).

[14] The result of the two-sided significance test of the HTMT relationship of the correlations is found in Appendix 5.2 of this chapter. The two-sided significance test was performed on the basis of a BCa bootstrap after (1987) with 5000 subsamples (auxiliary data sets).

[15] We carried out a repeat model estimation using the SmartPLS 3 by Ringle et al. (2015) without taking into factoring in both y_9 und y_{11} and using the identical algorithm and parameter settings, as described in Sect. 5.2.3. A description of the relevant results is omitted here.

Construct	Indicator	Convergence validity			Internal consistency reliability		Discriminant validity
		Factor loadings	Indicator reliability[16]	AEV	Composite reliability	Cronbach's alpha	
		≥.70	≥.50	≥.50	≥.60	≥.60	Is the HTMT criterion significantly smaller than 1?
Role model behavior	y_1	.813	.661	.707	.944	.931	Yes
	y_2	.818	.669				
	y_3	.868	.753				
	y_4	.868	.753				
	y_5	.827	.684				
	y_6	.814	.663				
	y_7	.877	.769				
Trans-parency	y_8	.763	.582	.580	.938	.927	Yes
	y_9	.663	.440				
	y_{10}	.702	.493				
	y_{11}	.684	.468				
	y_{12}	.864	.746				
	y_{13}	.824	.679				
	y_{14}	.721	.520				
	y_{15}	.754	.569				
	y_{16}	.743	.552				
	y_{17}	.806	.650				
	y_{18}	.826	.682				
Compliance culture	y_{19}	.829	.687	.763	.906	.844	Yes
	y_{20}	.920	.846				
	y_{21}	.869	.755				
Whistle-blower readiness	y_{22}	.794	.630	.677	.913	.882	Yes
	y_{23}	.777	.604				
	y_{24}	.845	.714				
	y_{25}	.875	.766				
	y_{26}	.819	.671				
Compliance risk	y_{27}	.932	.869	.819	.947	.926	Yes
	y_{28}	.923	.852				
	y_{29}	.904	.817				
	y_{30}	.858	.736				

Fig. 5.13 Evaluation results of testing the reflective measurement model

that all introduced measurement variables display a high explanatory power and that they measure the constructs in a reliabel and valid manner.

With respect to the formative measurement model, we first note that nearly all found outer weights are "appreciable" and lie above the recommended limit of ˃±0.10. Only the values of x_6 (−0.014) and x_8 (−0.063) fall below it. Nor do we detect any serious multicollinearity between the individual measured variables. The calculated variance inflation factor (VIF) values all clearly lie under the threshold of ≥5.0 that is widely accepted in the literature and for which, if exceeded, eliminating the impacted measurement variables, for example, is recommended. This can be evaluated as evidence of the robustness or efficiency of the estimated outer weights. In terms of the significance of the respective outer weights, a differentiated picture

Construct	Measurement variable	Outer weights	Outer loadings	VIF values	95 % BCa confidence intervall[a] (p = 0.05)	
		>±,10	>,50	≤5	Outer weights	Outer loadings
Incentives	x_1	-.130	.226	1.429	[-.381; .113]	[-.083; .530]
	x_2	.829	.900	2.203	[.489; 1.15]	[.804; .982]
	x_3	.520	.726	2.362	[.186; .778]	[.557; .883]
	x_4	-.121	.512	1.979	[-.427; .181]	[.213; .720]
	x_5	.122	.646	3.018	[-.221; .504]	[.438; .838]
	x_6	-.014	.657	2.791	[-.402; .332]	[.448; .828]
	x_7	-.226	.347	2.208	[-.587; .108]	[.071; .584]
	x_8	-.063	.318	1.384	[-.316; .175]	[.070; .588]

Fig. 5.14 Evaluation results of testing the formative measurement model. a. Two-sided significance test based on a BCa bootstrap after (1987) with 5000 subsamples (auxiliary data sets) (truncated illustration)

emerges. Thus, a two-sided significance test of the outer weights shows that solely the multiple regression coefficients of both x_2 and x_3 differ significantly from 0. However, this picture relativizes as soon as the found outer loadings are included in the evaluation. In that case, the result of a two-sided significance test of the outer loadings shows that all outer loadings—except for those of x_1 —differ significantly from 0. As for the extent of the outer loadings, it remains further that almost all the outer loadings exceed the recommended limit of ≥0.50. Only the values of x_1 (0.226), x_7 (0.374) and x_8 (0.318) lie below it. If all recommended evalutaion criteria are applied, it can be finally stated that no measurement variable needs to be eliminated. While neither the outer weight nor the outer loading of the x_1 measured variables differ statistically significantly from 0, nevertheless, x_1 makes a valuable contribution to the content validity of the construct "incentives", so that on purely statistical criteria grounds we refrain from eliminating it. Therefore, for the formative measurement model we can state that all deployed measurement variables exhibit a high explanatory power and that they measure the construct "incentives" in a reliabel and valid manner. We conclude therefore that the formal specification of the measurement models is empirically confirmed and we can proceed from a reliable and valid measurement of the constructs.

5.3.2.2 Evaluation of the Structural Model

For evaluating the structural model's estimated path coefficients (B and Γ) as well as its explanatory and predictive power, in what follows we address the result of testing the inner goodness of fit of the structural model using the goodness of fit criteria shown in Fig. 5.12. While the evaluation of the path coefficients normally takes place with collinearity (variance inflation factor) between the latent variables as well as their strength and significance, the evaluation of the explanatory and predictive power typically ensues using the determination coefficient, the f^2-strength, and the Stone-Geisser-criterion (Q^2).[16] These goodness of fit criteria are also automatically indicated along with the results report of SmartPLS 3 by Ringle et al. (2015) or can be calculated instantly with the software program based on the calculated results, so that here also conveniently any recalculation, such as with IBM SPSS or Microsoft Excel, becomes redundant.[17]

Thus, the predictor variables are largely independent of one another and no serious multicollinearity is discernible between them. The calculated VIF values all lie clearly below the threshold of ≥5.0 that is widely accepted in the literature.[18] This can be evaluated as proving the robustness or efficiency of the estimated path coefficients. Regarding the strength of these path coefficients, they make manifest that without exception we can assume "significant" relationships (≥0.20) between the constructs (Chin 1998b, p. 11). The result from a two-sided significant test of the path coefficients, moreover, shows that all estimated path coefficients differ significantly from 0. This is confirmed by the corresponding p-values.[19] Since the algebraic signs of the structural model estimation conform to the hypothetical structural model relationships and, furthermore, all path coefficients without exception differ significantly from 0, the assumed causal relationships between the latent variables cannot be discarded.

In terms of the indirect causal effects, it can be stated that these also without exception differ significantly from 0. This is confirmed by the corresponding p-values.[20] Thus, for example, the influence exerted by compliance culture on compliance risk is in part also significantly explained by employees' readiness to report rule violations. This "whistleblower readiness" therefore represent a mechanism that reveals an additional, relationship between "compliance culture" and "compliance risk" that works in the opposite direction. Thus, an improvement in the compliance culture, on the one hand, causes the compliance risk to drop, while, on the other hand, it also results in raising the readiness of employees to report rule violations, which once again raises the compliance risk. Should the "whistleblower readiness" therefore be boosted by an improved compliance culture, in the final analysis this suppresses *in part* the negative direct causal effect (−0.357), that an improved compliance culture exerts on compliance risk—and in an amount equal to the

[16] For an introduction to the goodness of fit criteria described in what follows for evaluating the structural model's path coefficients as well as its explanatory and prognostic power, see, for example, Hair et al. (2016, p. 190–209).

[17] The threshold values consulted in what follows are widely accepted benchmarks that are normally cited in the literature (Hair et al. 2016, p. 190–209).

[18] The calculated VIF values are found in Appendix 5.3 to this chapter.

[19] The two-sided significance test was performed based on a BCa bootstrap after Efron (1987) with 5000 subsamples (auxiliary data sets) and is found in Appendix 5.4 to this chapter. The related significance levels can be taken directly from depicted T-statistic.

[20] A two-sided significance test was performed based on a BCa bootstrap after Efron (1987) with 5000 subsamples (auxiliary data sets) and is found in Appendix 5.7 to this chapter. The related significance levels can be taken directly from depicted T-statistic.

positive indirect causal effect (0.138) mediated by whistleblower readiness. These kinds of effects are described as mediator effects (or interaction effects) (e.g., Hair et al. 2016; Schloderer et al. 2009; Baron and Kenny 1986) and were also to be expected against the background of the study concept.[21] If, in conclusion, the total causal effect are also included in the evaluation, it emerges that these as well differ significantly from 0. This is confirmed by the corresponding p-values.[22] Thus, compliance culture, for example, overall exerts a significant negative influence on the compliance risk, and does so in an amount of a negative total causal effect of −0.219—despite the suppressor effect brought to bear by the "whistleblower readiness."

Measured by the amount of the determination coefficient (R^2-values) the result moreover shows that "role model behavior" (0.369), "transparency" (0.477), "compliance culture" (0.456) and "whistleblower readiness" (0.240) consistently display good to very good values that are clearly higher than the threshold of ≥0.19 that is widely accepted in the literature (e.g., Chin 1998b, p. 325). Only "compliance risk" falls below it with a value of 0.109 This picture is confirmed by referring to the corrected determination coefficients (R^2- adjusted).[23] In terms of the effect strength (f^2-values) it is evident, moreover, that the exogenous constructs collectively influence the corresponding endogenous constructs significantly.[24] Overall, against the background of the study concept, the explanatory power of the structural model can be rated as very good. With respect to the concluding evaluation of the structural model's predictive relevance using the Stone-Geisser Q^2-value criterion, it emerges also that all the exogenous constructs with values clearly larger than 0 demonstrate predictive relevance with respect to their corresponding endogenous constructs and that the structural model overall possesses a very good predictive relevance.[25]

In conclusion, when all existing evaluation criteria are brought to bear on evaluating the path coefficients in the structural model as well as its explanatory and predictive power, the overall nomological validity of the theoretical and/or logically

[21] For analyzing the mediator effects in the structural model, we chose a systematic procedure after Hair et al. (2016, p. 233) that forms Appendix 5.9 to this chapter. For more detailed information on analyzing mediator effects, see for example the relevant introduction in Hair et al. (2016).

[22] A two-sided significance test was performed based on a BCa bootstrap after Efron (1987) with 5000 subsamples (auxiliary data sets) and is found in Appendix 5.8 to this chapter. The related significance levels can be taken directly from depicted T-statistic.

[23] Since the level of the determination coefficients is always also influenced by the number of regressors, Hair et al. (2016, p. 198 ff.) recommend always reverting to the corrected determination coefficients in evaluating the explanatory power of a structural model. Doing so the determination coefficients are corrected for the number of regressors. This then allows comparing the determination coefficients for different constructs, even if the numbers of assigned exogenous variables differ greatly. In the present actual use case, the corresponding values of the determination coefficients are 0.361 "role model behavior"), 0.464 "transparency"), 0.436 "compliance culture), 0.231 ("whistleblower readiness") and 0.087 ("compliance risk").

[24] The calculated f^2 values can be found in Appendix 5.5.

[25] The Q^2 values calculated using the blindfolding procedure are found in Appendix 5.6.

grounded causal relationships between the constructs—hence *also* that of the Compliance Index—is confirmed empirically, even given the background of the results of testing the outer goodness of fit of the measurement models.

5.3.3 The Compliance Index

The Compliance Index is a key performance indicator (KPI) of an organization's compliance culture. It shows wether a compliance program has been well-received by the employees, and is thus making a contribution towards mitigating compliance risks. The Compliance Index therefore enables quantitative, cost-efficient, internal and external benchmark comparisons with respect to the effectiveness of compliance programs. Despite the complexity of model estimation, the index is reasonably uncomplicated. The general form of the index is as follows:

$$\text{Compliance Index} = \frac{E[\eta_3] - \min[\eta_3]}{\max[\eta_3] - \min[\eta_3]} \times 100,$$

where η_3 , is the latent variable for the overall compliance culture, and *E* [.], *Min* [.] and *Max*[.] respectively denote the expected, the minimum, and the maximum value of the variable. The minimum and the maximum values are determined by those of the corresponding measurement variables:

$$\min[\eta_3] = \sum_{n}^{i=1} \pi_i \min[y_i].$$

and

$$\max[\eta_3] = \sum_{n}^{i=1} \pi_i \max[y_i]$$

where y_i's are the measurement variables of the latent variable compliance culture, π_i's are the outer weights, and *n* is the number of measurement variables. In calculating the Compliance Index, unstandardized outer weights must be used if unstandardized measurement variables are used.

In the Compliance Index Model, the index score is calculated as a weighted average of three indicators that range from 1 to 10. Then, the calculation is simplified to

$$\text{Compliance Index} = \frac{\sum_{i=1}^{3} \pi_i y_i - \sum_{i=1}^{3} \pi_i}{9 \sum_{i=1}^{3} \pi_i} \times 100,$$

where the π_i's are the unstandardized outer weights.

5.4 Importance-Performance-Map Analysis (IPMA)

While the primer focuses on Compliance-Index-Modell foundations, this chapter is on more advanced analytics that complement the basic Compliance-Index-Modell. IPMA is a useful approach in the Compliance Index Model that enables firms to leverage the respected, predictive capabilities of the Compliance Index Model and obtain detailed, actionable insights for improving an organization's compliance culture: first, users can determine which constructs and indicators are most critical for driving compliance culture and, consequently, mitigating compliance risks. Second, users obtain strategic and practical insights about the drivers of compliance culture from the perspective of its relationship to compliance risks. Finally, users can develop data-driven action plans to improve an organization's compliance culture and, consequently, mitigate compliance risks.[26]

5.4.1 *Step 1: Computation of the Performance Values*

The indicator data determines the latent variable scores and, thus, their performance. Similarly, when conducting an IPMA on the indicator level, the mean value of an indicator represents its average performance. To facilitate the interpretation and comparison of performance levels, the IPMA rescales indicator scores on a range between 0 and 100, with 0 representing the lowest and 100 representing the highest performance. Since most researchers are familiar with interpreting percentage values, this kind of performance scale is easy to understand (e.g., Ringle and Sarstedt 2016, p. 1869). In the Compliance Index Model, the rescaling of an observation *j* with respect to indicator *i* proceeds via.

$$y_{ij}^{rescaled} = \frac{E\left[y_{ij}\right] - \min\left\lfloor y_{ij} \right\rfloor}{\max\left[y_i\right] - \min\left[y_i\right]} \times 100,$$

[26] Generally, IPMA is a useful approach in PLS that extends the standard results reporting of path coefficient estimates by adding a dimension that considers the average values of the latent variable scores. More precisely, the IPMA contrasts the total effects, representing the predecessor constructs' importance in shaping a certain target construct (such as "compliance culture"), with their average latent variable scores indicating their performance, both in an unstandardized form (e.g., Hair et al. 2018; Hair et al. 2016; Ringel and Sarstedt 2016; Fornell et al. 1996). The goal is to identify predecessors that have a relatively high importance for the target construct (i.e., those that have a strong total effect), but also have a relatively low performance (i.e., low average latent variable scores). By combining the analysis of the importance and performance dimensions, the IPMA allows for prioritizing constructs to improve the target construct. Expanding the analysis to the indicator level allows for identifying the most important areas of specific actions. The description that follows is based on Ringel and Sarstedt (2016), who offer a relevant, practical introduction to IPMA. See also Hair et al. (2018) and Hair et al. (2016) for further information, especially within in the context of SmartPLS 3 (Ringle et al. 2015).

Case	x_1	x_2	x_3	...	y_4	y_5	y_6	y_7
1	4	8	9	...	3	4	7	8
2	5	2	4	...	7	8	8	8
3	2	1	1	...	9	10	9	9
4	3	8	5	...	5	6	6	5
5	8	5	1	...	8	10	10	5
...	...	...	...	...	...	...	...	...
84	6	7	8	...	4	6	4	6
85	4	2	3	...	10	9	8	8
Mean value	4.7	4.9	5.6	...	5.5	6.6	6.5	6.5

Fig. 5.15 Original indicator data

Case	x_1	x_2	x_3	...	y_4	y_5	y_6	y_7
1	67	22	11	...	22	33	67	78
2	56	89	67	...	67	78	78	78
3	89	100	100	...	89	100	89	89
4	78	22	56	...	44	56	56	44
5	22	56	100	...	78	100	100	44
...	...	...	...	...	...	...	...	...
84	44	33	22	...	33	56	33	56
85	67	89	78	...	100	89	78	78
Mean value	62.1	56.5	52.9	...	50.3	63.8	62.0	61.7

Fig. 5.16 Rescaled indicator data

and

$$x_{ij}^{rescaled} = \frac{E\left[x_{ij}\right] - \min\left[x_{ij}\right]}{\max\left[x_i\right] - \min\left[x_i\right]} \times 100,$$

where y_i and x_i are the ith indicator, respectively, E[.] represents indicator i's actual score of respondent j, and *Min* [.] and *Max*[.] represent the indicator's minimum and maximum value. It is important to note that the minimum and maximum values refer to the potential values on the ten-point scale and not the minimum and maximum values of the actual responses (e.g., 4 on a ten-point scale). Hence, if the respondents' lowest actual response is 4 but the scale has a minimum value of 1, it is mandatory to use 1 as minimum value for rescaling. For example, according to this formula, a value of 4 on a ten-point scale becomes $(4 - 1)/(10 - 1) \times 100 \approx 33$.

Figure 5.15 shows an excerpt of the original indicator data ($N = 85$) used to estimate the Compliance Index Model. All indicators are measured on a ten-point scale. Figure 5.16 shows the indicator data from Fig. 5.15, rescaled on a range from 0 to

100, which serve as input for the computation of the rescaled latent variable scores.[27] In addition, the mean values of the rescaled indicators represent their performance values (e.g., 67 for x_1, and 22 for x_2), which are later used for the IPMA on the indicator level.

The rescaled latent variable scores are a linear combination of the rescaled indicator data and the rescaled outer weights—regardless whether the measurement model of a latent variable is reflective or formative (e.g., Ringle and Sarstedt 2016, p. 1870). To obtain the rescaled outer weights, we must first compute the unstandardized outer weights by dividing the standardized outer weights by the standard deviation (SD) of its respective indicator. While the standardized outer weights originate from the Compliance Index Model estimation, the estimation of each indicator's standard deviation is based on the original indicator data.[28] For example, if x_1 has a standardized weight of 0.130 and a standard deviation of 2.714, the resulting unstandardized weight is 0.048. Figure 5.17 shows the (standardized and unstandardized) outer weights along with the indicators' standard deviations with regard to the direct predecessors of the selected target construct "compliance culture".

Finally, we rescale the unstandardized outer weights so that their sum equals one per measurement model. For this purpose, we need to divide each indicator's unstandardized regression weight (e.g., 0.048 for x_1) by the sum of the unstandardized weights of all the indicators that belong to the same construct. For instance, for the construct "incentives", the sum of all the unstandardized weights is 0.062 + 0.324 + 0.185 + 0.044 + 0.048) + 0.005 + 0.086 + 0.029 = 0.768. Therefore, for x_1, we obtain the unstandardized and rescaled outer weight of 0.062 after dividing 0.048 by 0.768. The final column in Fig. 5.17 shows the results of the rescaled outer weights.

In the next step, the IPMA uses the rescaled indicator data (Fig. 5.16) and the rescaled outer weights (Fig. 5.17) to compute the rescaled latent variable scores by means of simple linear combinations. For example, the first data point in the vector of the construct incentives's score is $67 \times 0.062 + 22 \times 0.422 + 11 \times 0.240\ldots \approx 26.6$.

[27] In an IPMA, all the indicator coding must have the same scale direction (Ringle and Sarstedt 2016, p. 1868). The minimum value of an indicator must represent the worst outcome and the maximum value must represent the best outcome of an indicator. Otherwise, we cannot conclude that higher latent variable scores represent better performance. If the indicator coding has a different direction compared to the other indicators in the measurement model (i.e., a high value represents a negative outcome), we must rescale the indicator. Hence, the indicator coding of the construct "incentives" has been changed by reversing the scale (e.g., on a ten-point scale, 10 becomes 1 and 1 becomes 10, 8 becomes 4 and 4 becomes 8, and 5 remains unchanged).

[28] Regardless of the measurement model being formatively or reflectively specified, the outer weights must be positive in an IPMA (Ringle and Sarstedt 2016, p. 1868). If the outer weights are negative, the latent variable score will not fall within the 0–100 range, but would, for example, be between −10 and 90. Hence, in case of the construct "incentives" all negative outer standardized weights have been changed by reversing the signs.

Latent variable	Indicator	Outer weights (standardized)	Standard deviation	Outer weights (unstandardized)	Outer weights (rescaled)
Incentives	x_1	.130	2.714	.048	.062
	x_2	.829	2.558	.324	.422
	x_3	.520	2.818	.185	.240
	x_4	.121	2.768	.044	.057
	x_5	.122	2.560	.048	.062
	x_6	.014	2.785	.005	.007
	x_7	.226	2.631	.086	.112
	x_8	.063	2.141	.029	.038
Trans-parency	y_9	.145	1.920	.076	.082
	y_{10}	.128	1.953	.066	.057
	y_{11}	.143	2.181	.066	.067
	y_{12}	.098	1.887	.052	.079
	y_{13}	.073	2.018	.036	.098
	y_{14}	.087	2.052	.042	.104
	y_{15}	.105	2.104	.050	.080
	y_{16}	.145	2.335	.062	.105
	y_{17}	.142	2.151	.066	.119
	y_{18}	.104	2.044	.051	.104
	y_{19}	.128	1.936	.066	.104
Role-model behavior	y_1	.161	1.848	.087	.159
	y_2	.145	2.188	.066	.121
	y_3	.167	2.067	.081	.148
	y_4	.207	2.399	.086	.158
	y_5	.190	2.371	.080	.147
	y_6	.147	2.267	.065	.119
	y_7	.170	2.095	.081	.148

Fig. 5.17 Unstandardized and rescaled outer weights

Case	Incentives	Transparency	Role-model behavior
1	26.6	41.3	56.5
2	76.3	71.7	67.7
3	98.0	82.0	90.2
4	47.3	42.6	50.4
5	69.0	57.3	70.6
...	...	...	...
84	31.3	47.3	51.2
85	77.1	70.1	87.3
Mean value	56.8	56.7	60.3

Fig. 5.18 Rescaled construct values

Figure 5.18 shows the resulting latent variable scores along with their mean values. In the Compliance Index Model, the predecessors of the selected target construct "compliance culture" have a mean value (i.e., performance) of 56.8 ("incentives"), 56.7 ("transparency"), and of 60.3 ("role-model behavior"). These results serve as input for the importance-performance map's performance dimension.

Latent variable	Direct effect	Indirect effect	Total effect
Incentives	-.147	-.062	-.209
Transparency	.246		.246
Role-model behavior	.209	.265	.474

Fig. 5.19 Direct, indirect and total effects in the IPMA

5.4.2 Step 2: Computation of the Importance Values

A construct's importance in terms of predicting another directly or indirectly linked (target) construct in the structural model is derived from the total effect of the relationship between these two constructs. Since total effects represent the sum of direct and indirect effects, the IPMA's importance dimension supports the interpretation of complex models including meditators or even multiple mediators (e.g., Ringle and Sarstedt 2016, p. 1872).

The IPMA draws on unstandardized effects to facilitate a ceteris paribus interpretation of predecessor constructs' impact on the target construct. This interpretation of the unstandardized effects is analogous to that of unstandardized weights in OLS regression models (Hair et al. 2016). More precisely, by drawing on unstandardized effects, we can conclude that an increase in a certain predecessor construct's performance would increase the target construct's performance by the size of its unstandardized total effect (e.g., Ringle and Sarstedt 2016, p. 1872).

Figure 5.19 summarizes all unstandardized effects with respect to the selected target construct "compliance culture". Note that the construct "transparency" does not have an indirect effect on "compliance culture"; therefore, its total effect equals the direct effect of 0.246. At this point, after computing the importance and performance values, all information required to draw the importance-performance map is available.

5.4.3 Step 3: Importance-Performance Map Creation

An attractive feature of the importance-performance map analysis is that the results may be graphically displayed on an easily-interpreted, two-dimensional map. Figure 5.20 summarizes the values of this map's two dimensions—as obtained by the previous IMPA steps. In case of the construct "incentives" the importance value has been adjusted by reversing the sign.

Scatter plotting the information shown in Fig. 5.20 allows us to create the importance-performance map for the selected target construct "compliance culture".

	Importance	Performance
Incentives	.209	56.8
Transparency	.246	56.7
Role-model behavior	.474	60.3
Mean value	.310	57.9

Fig. 5.20 IPMA data for the construct "compliance culture"

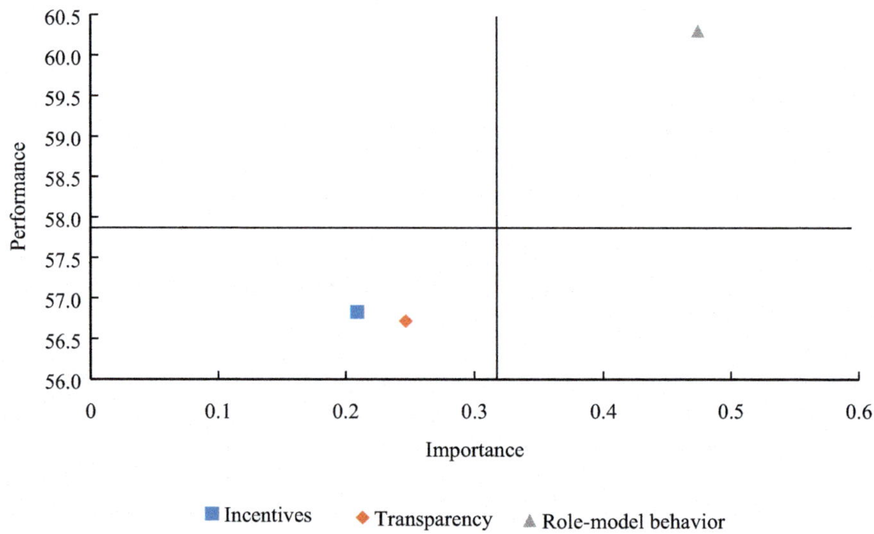

Fig. 5.21 IMPA on the construct "compliance culture" (construct level)

The x-axis represents the importance of the constructs "incentives", "transparency", and "role-model behavior" for explaining "compliance culture", while the y-axis depicts their performance in terms of their average rescaled latent variable scores. For a better orientation, two additional lines have been drawn in the importance-performance map: the mean importance value (i.e., a vertical line) and the mean performance value (i.e., a horizontal line) of the displayed constructs (Fig. 5.21).

In the Compliance Index Modell, the constructs "incentives", "transparency", and "role-model behavior" have a mean importance of 0.310 and a mean performance of 57.9 (Fig. 5.20). These two additional lines divide the importance-performance map into four areas with importance and performance values below and above the average. Generally, when analyzing the importance-performance map, constructs in the lower right area (i.e., above average importance and below average performance) are of highest interest to achieve improvement, followed by the higher right, lower left and, finally, the higher left areas. Thereby, the importance-performance map provides guidance for the prioritization of managerial activities of high importance for the aspects underlying the selected target construct, but which require performance improvements.

Not surprisingly, we find that "role-model behavior" has a relatively high performance (Fig. 5.21). In comparison with the other constructs, "role-model behavior"'s performance is slightly above average. Further, with a total effect of 0.474, this construct's importance is particularly high. Therefore, a one-unit increase in "role-model behavior"'s performance would increase the performance of "compliance culture" by 0.474 points. Hence, when managers aim at increasing the performance of the selected target construct "compliance culture", their first priority should be to improve the performance of aspects captured by "role-model behavior", as this construct has the highest (above average) importance and a relatively high (above average) performance. Aspects related to the constructs "transparency" and "incentives" follow as a second and third priority. It is important to note that different data input would produce different IPMA-results for the selected target construct "compliance culture" (e.g., changing the constructs' priority order).

The IPMA is not limited to the construct level. We can also conduct an IPMA on the indicator level to identify relevant and even more specific areas of improvement. More precisely, we can interpret the rescaled outer weights—as reported in formative measurement models—as an indicator's relative importance compared to that of the other indicators in a specific measurement model (e.g., Ringle and Sarstedt 2016, p. 1874). Alternatively, the interpretation of the indicators' relative contribution can also draw on reflective measurement models but use the outer weights instead of the outer loadings (2016, p. 1874). While the outer weights play no role in the assessment of the reflective measurement model's reliability and validity, they still represent each indicator's contribution to forming the composite variables that represent the constructs in the Compliance Index Modell.

The importance values are derived from the indicators' total effects on the selected target construct "compliance culture", which is the result of multiplying the rescaled outer weights of a predecessor construct's indicators with its unstandardized total effect on "compliance culture". For example, with regard to the indicators of "role-model behavior", we would multiply the rescaled outer weights of y_1-y_7 (i.e., 0.159, 0.121, 0.148, 0.158, 0.147, 0.119, and 0.148; Fig. 5.17) with the unstandardized total effect of "role-model behavior" on "compliance culture" in the structural model (i.e., 0.474). This analysis yields importance values of y_1–y_7 of, respectively, 0.076, 0.057, 0.070, 0.075, 0.069, 0.056, and 0.070. The performance values are derived from the indicators' mean value of the rescaled data (Fig. 5.16). With this data for all indicators of the constructs "incentives", "transparency", and "role-model behavior", we can create an importance-performance map as shown in Fig. 5.22. In this figure, rectangles represent the indicators of the construct "incentives", diamonds those of the construct "transparency", and triangles those of the construct "role-model behavior".

Derived from this analysis, we find that indicator y_4 ("I am convinced that the chief executive and the rest of the management team have a higher regard for ethics

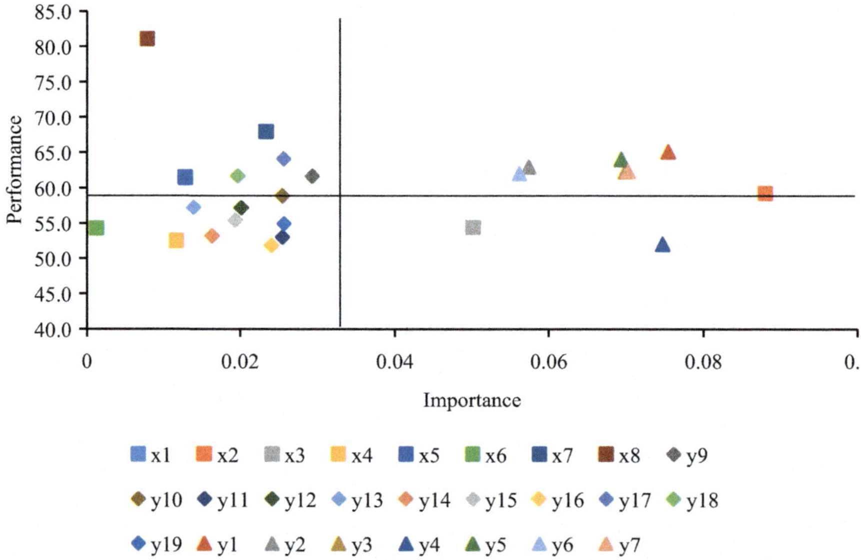

Fig. 5.22 IPMA on the construct "compliance culture" (indicator level)

and values than short-term business targets.") for "role-model behavior" should have the highest priority for improvement, since it has the highest relative importance, but the lowest performance. Hence, performance improvements may focus on ethics trainings to motivate top management to "walk the talk" in terms of behaving consistently with the firm's standards of conduct, which articulate an organization's expectations regarding ethics and values. As a direct consequence, the performance of the construct "role-model behavior" increases, which entails an improvement of the construct "transparency", and the selected target construct "compliance culture". Similarly, other indicators (e.g., indicator x_3 for "incentives ""I am convinced that fellow employees and managers violate existing standards of conduct because they feel pressured to achieve business targets at all costs.") may gain particular attention regarding improving „compliance culture". Again, it is important to note that different data input would produce different IPMA-results for the selected target construct "compliance culture" (e.g., changing the indicators' priority order).

5.5 Discussion and Conclusion

5.5.1 Discussion

An extensive summary of research on the effectiveness of compliance programs has stated that the scope and variety of elements (such as a code of conduct, a compliance officer, a training concept as well as monitoring and control measures) in such programs are of more importance than employees' perceptions of the programs (e.g., Kaptein 2014). Our study stands in contrast to this theoretical stream in two important ways: by demonstrating that incentives, role-model-behavior, and transparency are factors that determine how compliance programs come to be perceived in an organization, and by highlighting compliance culture as a determinant of whistleblower readiness and compliance risk. For example, a poor compliance culture may be more likely to lead to misconduct when existing organizational reward structures create a priori incentives for corner-cutting behaviors, and when mechanisms exist to easily facilitate the diffusion of shared perceptions and of system-beating practices.

Our findings imply that it is insufficient to discuss how many or what types of formal elements characterize a compliance program, because programs are likely to be ineffective in a poor compliance culture, regardless of their scope. To achieve desired outcomes, concerns for compliance must be baked into the culture of the organization. Therefore, attention to the compliance culture should come first in any corporate compliance effort. Executive leaders and supervisors must regularly show they care about compliance, and they must show that they care through words and consistent actions.

Our findings further suggest that responsibility for compliance management should be held broadly within the organization. If one individual is ultimately responsible, that individual should be a high-level corporate officer with credibility, line management experience, a strong belief in the power of organizational values, solid support from the organization's executive leadership, and the ability to work with many other officers and departments (such as legal counsel, human resources, audit) toward the development of a strong compliance culture. In addition, human resources should have an expanded role in compliance management because of the importance of issues such as employee fair treatment.

Finally, future research could conduct multigroup analyses to test if pre-defined data groups at multiple levels (such as industry, organization, and unit) have significant differences in their group-specific Compliance Index Model parameter estimates (e.g., outer weights, outer loadings, and path coefficients). SmartPLS 3 by Ringle et al. (2015) provides outcomes of three different approaches that are based on bootstrapping results from every group. E.g., Hair et al. (2016, 2018) describe the multigroup analysis methods in detail.

5.5.2 Conclusion

We expect the Compliance Index Model to be generally applicable to banks and other financial institutions. The model and measures are designed to provide this general applicability. As such, they represent a significant step forward in evaluating—and enhancing—the effectiveness of compliance programs. The model provides an independent and uniform tool for obtaining strategic and practical insights into the key drivers of an organization's compliance culture from the perspective of its relationship to compliance risks. Looking at the model variables and effects, users can determine which drivers of compliance culture to improve for the most impact on compliance risks. Moreover, users can compare their results with industry peers and competitors and also perform cross-industry benchmarking with companies they deem "best-in-class" in other industries to reveal critical gaps. Consequently, compliance culture can be actively shaped, monitored, and sustained by committed leaders and organizations to mitigate compliance risks.

For public policymakers, the Compliance Index Model is a potentially useful tool for evaluating and enhancing the safety and soundness of the banking system, managing financial integration, enhancing banking system stability and providing consistent supervision. As such, it can furnish an important complement to conventional oversight measures pertaining to banks and other financial institutions—in conducting supervisory reviews, on-site inspections and investigations—and it can balance these measures against ongoing initiatives designed to rebuild trust in the banking sector and to increase the resilience of banks. In the legislative arena, the model can help to predict and monitor the effects of public policy decisions and ensure the consistent application of the regulations and supervisory policies they spawn. In terms of compliance risk, it can provide an early warning as to whether a bank or other financial institution is vulnerable to misconduct.

For managers and investors, the Compliance Index Model provides an important measure of the institution's past and current compliance program effectiveness as well as its compliance risk exposure. It furnishes a means for measuring one of the most fundamental prerequisites for an institution's compliance program's effectiveness—its compliance culture. A stronger compliance culture increases whistleblower readiness and mitigates compliance risks. As such, the model supplies a leading indicator of an institution's current and future financial health. By establishing a standard measure of the overall compliance culture with clear links to long-term performance, the model may even help instill more of a long-term perspective in both management and investors.

The Compliance-Index-Modell also has implications for managers formulating competitive strategy. One of its key benefits is that the model represents a uniform and comparable system of measurement that allows for systematic benchmarking over time and across firms. In addition, it can be useful in analyzing the strengths and weaknesses of the firm or its competitors. For example, declining compliance culture is likely to be symptomatic of deeper problems facing a firm. Because the model provides a measure of the effectiveness of a firm's compliance program, firms with low Compliance Index scores are particularly vulnerable to misconduct and compliance risk; thus provide no expansion opportunities for more competent organizations.

For customers, the Compliance Index Model provides information that is not only useful in strengthening consumer rights but also likely to help improve the quality of the banking products and services they consume. The independence, uniformity, and methodology built into the model lets customers access information that is not available from rating agencies and commercial market research firms and is missing from their ad hoc methodologies. Arguably, the mere availability of a tool like the Compliance Index Model may spur enhancements to consumer rights and improve the quality of banking products and services.

To conclude, the Compliance Index Model represents a new tool for evaluating and enhancing the perceived effectiveness of compliance programs. It produces critical information to fill a gap in the knowledge we need to ensure that compliance programs address the root causes of compliance lapses rather than just "treating the symptoms". It thus has the potential to significantly enhance the effectiveness of compliance programs in banks and other financial institutions striving to survive in an increasingly challenging marketplace. Banking and finance scholars and practitioners have long recognized compliance culture as a key concept for successfully managing businesses of all kinds. The Compliance Index Model has the potential for strengthening compliance culture, mitigating the attendant risks and maximizing its beneficial effects.

Acknowledgement This research was financially supported by FIRM (Frankfurt Institute of Risk Management and Regulation; www.firm.fm).

Appendix 5.1

Operationalized latent variables of the structural model

Question (measured indicator)	Measured variable
Incentives (ξ_1)	*Formative*
On a scale from "1 (very little)" to "10 (very strongly)," how convinced are you that fellow employees and managers in your firm violate existing standards of conduct because they ...	
...either do not understand them or are not familiar enough with them?	x_1
...believe that the firm's code of conduct is not meant to be taken seriously?	x_2
...feel pressured to achieve business targets at all costs?	x_3
...are not provided with sufficient resources (for example, time, budget, equipment, information, and authorities) for performing their daily duties correctly?	x_4
...believe that guidelines and procedure manuals can easily be circumvented or overridden?	x_5
...believe that they are rewarded only for their work results and not for the manner in which they achieved them?,	x_6
...fear that they might lose their job if they fail to meet their goals?	x_7
...want to enrich themselves personally?	x_8
Role model behavior (η_1)	*Reflective*
On a scale from "1 (very little)" to "10 (very strongly)," how convinced are you that the chief executive and the rest of the management team ...	
...live up to their function as role models for the firm?	y_1
...are amenable to listening to concerns voiced by employees about conduct that they regard as illegal, unethical, or at least questionable?	y_2

(continued)

(continued)

Question (measured indicator)	Measured variable
…can be approached with bad news conveyed to them by their employees?	y_3
…have a higher regard for ethics and values than short-term business targets?	y_4
…set targets that can be achieved also without violating the firms's code of conduct?	y_5
…would react appropriately in case they find out about misconduct?	y_6
…make a believable case for which behavior is acceptable in the firm and which is not? ("the tone is set at the top")	y_7
Transparency (η_2)	*Reflective*
On a scale from "1 (very little)" to "10 (very strongly)," how convinced are you that in your firm people are careful when implementing measures and instruments to …	
…detect misconduct?	y_8
…be conscientious about following the rules?	y_9
…follow up on complaints?	y_{10}
…enforce guidelines and standards??	y_{11}
…check employee conduct?	y_{12}
…penalize rule violations?	y_{13}
…inform and counsel employees?	y_{14}
…promote ethically unobjectionable behavior and integrity?	y_{15}
…foster shared values?	y_{16}
…evaluate conformance with rules (or compliance)?	y_{17}
…help employees make decisions?	y_{18}
Compliance culture (η_3)	*Reflective*
On a scale from "1 (very little)" to "10 (very strongly)," how convinced are you that …	
…your colleagues are by and large open in voicing their concerns regarding illegal, unethical, or at least questionable conduct?	y_{19}
…your colleagues in making decisions and in conducting themselves by and large adhere to the right regulations and standards?	y_{20}
…your colleagues by and large feel obligated to conduct themselves in accordance with the rules?	y_{21}
Whistleblower readiness (η_4)	*Reflective*
Should you report a violation of the firm's prevailing standards of conduct, how likely, on a scale from "1 (not at all)" and "10 (very sure)," do you think it is that	
…there will be an appropriate response?	y_{22}
…your report would be treated confidentially?	y_{23}
…you are protected against retaliation or reprisals?	y_{24}
…those involved in the violation are sanctioned appropriately without consideration given to their professional rank in the firm?	y_{25}
…you would be satisfied with the consequences of your action?	y_{26}

(continued)

(continued)

Question (measured indicator)	Measured variable
Compliance risk (η_5)	*Reflective*
How likely, on a scale from "1 (not at all)" and "10 (very sure)," do you think it is that the violations observed by you, should they be revealed, would ...	
...result in a significant (public) loss of trust and reputation?	y_{27}
...lead to significant legal penalties or sanctions?	y_{28}
...cause a significant loss of new or existing customers?	y_{29}
...result in significant drop in employee morale or productivity?	y_{30}

Appendix 5.2

Result of the two-sided significance test of the HTMT relationship of correlations

BCa bootstrap result for HTMT[a]					
	Original sample (O)	Sample mean (M)	Bias	0.5%	99.5%
Compliance risk (η_5) → Compliance culture (η_3)	.245	.257	.012	.061	.467
Whistleblower readiness (η_4) → Compliance culture (η_3)	.550	.544	−.006	.259	.775
Whistleblower readiness (η_4) → Compliance risk (η_5)	.108	.162	.054	.046	.194
Transparency (η_2) → Compliance culture (η_3)	.640	.639	−.001	.377	.829
Transparency (η_2) → Compliance risk (η_5)	.105	.169	.064	.082	.114
Transparency (η_2) → Whistleblower readiness (η_4)	.439	.453	.014	.224	.661
Role model behavior (η_1) → Compliance culture (η_3)	.631	.628	−.003	.386	.800
Role model behavior (η_1) → Compliance risk (η_5)	.107	.170	.063	.045	.173
Role model behavior (η_1) → Whistleblower readiness (η_4)	.724	.721	−.003	.513	.861
Role model behavior (η_1) → Transparency (η_2)	.588	.591	.003	.364	.776

[a]Significance level $\alpha = 0.01$

Appendix 5.3

Collinearity statistic (VIF) of the predictor variables

VIF values[a]						
	Incentives (ξ_1)	Compliance culture (η_3)	Compliance risk (η_5)	Whistleblower readiness (η_4)	Transparency (η_2)	Role model behavior (η_1)
Incentives (ξ_1)		2.003			1.585	1000
Compliance culture (η_3)			1.315	1.000		
Compliance risk (η_5)						
Whistleblower readiness (η_4)			1.315			
Transparency (η_2)		1.912				
Role model behavior (η_1)		1.755			1.585	

[a]Critical VIF-value 5.0

Appendix 5.4

Result of the two-sided significance test of path coefficients

BCa bootstrap result for the path coefficients					
	Original sample (O)	Sample mean (M)	Standard-deviation (STABW)	T-Statistic[a] (|O/STABW|)	p-values
Incentives (ξ_1) → Compliance culture (η_3)	−.280	−.309	.109	2.565	.010
Incentives (ξ_1) → Transparency (η_2)	−.468	−.497	.110	4.244	.000
Incentives (ξ_1) → Role model behavior (η_1)	−.607	−.635	.071	8.608	.000
Compliance culture (η_3) → Compliance risk (η_5)	−.357	−.377	.099	3.618	.000
Compliance culture (η_3) → Whistleblower readiness (η_4)	.490	.498	.089	5.521	.000
Whistleblower readiness (η_4) → Compliance risk (η_5)	.283	.293	.115	2.461	.014
Transparency (η_2) → Compliance culture (η_3)	.254	.248	.103	2.466	.014
Role model behavior (η_1) → Compliance culture (η_3)	.250	.227	.104	2.397	.017
Role model behavior (η_1) → Transparency (η_2)	.298	.275	.117	2.552	.011

[a]Critical t-values 2.57 ($\alpha = 0.01$), 1.96 ($\alpha = 0.05$) and 1.65 ($\alpha = 0.10$) for two-sided significance test

Appendix 5.5

Result of the calculated effect strengths of the exogenous constructs

Effect strengths (f^2-values) of the exogenous constructs[a]

	Incentives (ξ_1)	Compliance culture (η_3)	Compliance risk (η_5)	Whistleblower readiness (η_4)	Transparency (η_2)	Role model behavior (η_1)
Incentives (ξ_1)		.072			.264	.585
Compliance culture (η_3)			.109	.315		
Compliance risk (η_5)						
Whistleblower readiness (η_4)			.068			
Transparency (η_2)		.062				
Role model behavior (η_1)		.065			.107	

[a]For evaluating effect strength (f^2-values) of the exogenous constructs, Chin (1998a, p. 317) proposes the following guideline values: 0.02 (small), 0.15 (medium) and 0.35 (high)

Appendix 5.6

Result of the calculated prognostic relevance of endogenous constructs

Cross validated redundancy for constructs

	SSO	SSE	Q^2 (=1-SSE/SSO)[a]
Incentives (ξ_1)	680	680	
Compliance culture (η_3)	255	174.633	0.315
Compliance risk (η_5)	340	314.029	0.076
Whistleblower readiness (η_4)	425	365.506	0.140
Transparency (η_2)	935	709.859	0.241
Role model behavior (η_1)	595	455.749	0.234

[a]Critical Q^2-value 0

Appendix 5.7

Result of the two-sided significance test of indirect causal effects

BCa bootstrap result of indirect causal effects

	Original sample (O)	Sample mean (M)	Standard-deviation (STABW)	T-Statistic[a] (|O/STABW|)	p-values
Incentives (ξ_1) → Compliance culture (η_3)	−.316	−.311	.088	3.584	.000
Incentives (ξ_1) → Compliance risk (η_5)	.131	.141	.059	2.221	.026
Incentives (ξ_1) → Whistleblower readiness (η_4)	−.292	−.310	.074	3.927	.000
Incentives (ξ_1) → Transparency (η_2)	−.181	−.171	.074	2.449	.014
Compliance culture (η_3) → Compliance risk (η_5)	.138	.146	.068	2.025	.043
Transparency (η_2) → Compliance risk (η_5)	−.056	−.055	.031	1.787	.074
Transparency (η_2) → Whistleblower readiness (η_4)	.124	.124	.058	2.154	.031
Role model behavior (η_1) → Compliance culture (η_3)	.076	.068	.043	1.742	.082
Role model behavior (η_1) → Compliance risk (η_5)	−.071	−.067	.036	1.972	.049
Role model behavior (η_1) → Whistleblower readiness (η_4)	.159	.150	.062	2.569	.010
Role model behavior (η_1) → Transparency (η_2)	−.316	−.311	.088	3.584	.000

[a]Critical t-values 2.57 ($\alpha = 0.01$), 1.96 ($\alpha = 0.05$) and 1.65 ($\alpha = 0.10$) for two-sided significance test

Appendix 5.8

Result of the two-sided significance test of total causal effects

BCa bootstrap result of t total causal effects

	Original sample (O)	Sample mean (M)	Standard-deviation (STABW)	T-Statistic[a] (|O/STABW|)	p-values
Incentives (ξ_1) → Compliance culture (η_3)	−.596	−.619	.070	8.506	.000
Incentives (ξ_1) → Compliance risk (η_5)	.131	.141	.059	2.221	.026
Incentives (ξ_1) → Whistleblower readiness (η_4)	−.292	−.310	.074	3.927	.000
Incentives (ξ_1) → Transparency (η_2)	−.649	−.670	.067	9.733	.000
Incentives (ξ_1) → Role model behavior (η_1)	−.607	−.634	.070	8.621	.000
Compliance culture (η_3) → Compliance risk (η_5)	−.219	−.227	.090	2.425	.015
Compliance culture (η_3) → Whistleblower readiness (η_4)	.490	.497	.087	5.618	.000
Whistleblower readiness (η_4) → Compliance risk (η_5)	.283	.291	.118	2.395	.017
Transparency (η_2) → Compliance culture (η_4)	.254	.246	.102	2.480	.013
Transparency (η_2) → Compliance risk (η_5)	−.056	−.055	.031	1.787	.074
Transparency (η_2) → Whistleblower readiness (η_4)	.124	.124	.058	2.154	.031
Role model behavior (η_1) → Compliance culture (η_3)	.326	.296	.099	3.283	.001
Role model behavior (η_1) → Compliance risk (η_5)	−.071	−.067	.036	1.972	.049
Role model behavior (η_1) → Whistleblower readiness (η_4)	.159	.150	.062	2.569	.010
Role model behavior (η_1) → Transparency (η_2)	.298	.272	.116	2.569	.010

[a]Critical t-values 2.57 ($\alpha = 0.01$), 1.96 ($\alpha = 0.05$) and 1.65 ($\alpha = 0.10$) for two-sided significance test

Appendix 5.9

Systematic approach to analyzing mediator effects

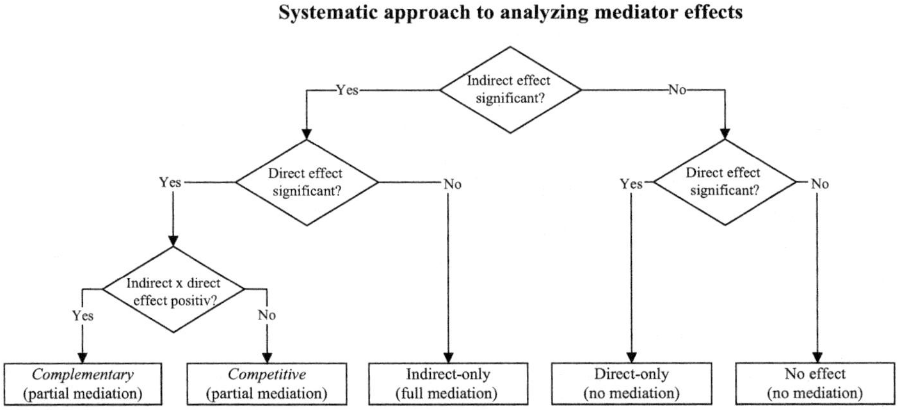

References

Allison, P. D. (2001). *Missing data (quantitative applications in the social sciences)*. Thousand Oaks: Sage.

Barker, R. (1993). An evaluation of the ethics program at General dynamics. *Journal of Business Ethics, 12*, 165–177.

Barladi, A. N., & Enders, C. K. (2010). An Introduction to modern missing data analyses. *Journal of School Psychology, 48*(1), 5–37.

Baron, R., & Kenny, D. (1986). The moderator-mediator variable distinction in social psychological research: Conceptual, strategic, and statistical consideration. *Journal of Personality and Social Psychology, 51*(6), 1173–1182.

Beeri, I., Dayan, R., Vigoda-Gadot, E., & Werner, S. (2013). Advancing ethics in public organizations: The impact of an ethics program on employees' perceptions and behaviors in a regional council. *Journal of Business Ethics, 112*(1), 59–78.

Biemer, P. P., & Lyberg, L. (2003). *Introduction to survey quality*. New York: Wiley.

Blodgett, M. S. (2011). Substantive ethics: Integrating law and ethics in corporate ethics programs. *Journal of Business Ethics, 99*(1), 39–48.

Chin, W. W. (1998a). The partial least squares approach to structural equation modeling. *Modern Business Research Methods, 295*, 295–336.

Chin, W. W. (1998b). Issues and opinion on structural equation modeling. *Management Information Systems Quarterly, 22*(1), 7–16.

Cronbach, L. J., & Meehl, P. E. (1955). Construct validity in psychological tests. *Psychological Bulletin, 52*(4), 281–302.

Delaney, J. T., & Sockell, D. (1992). Do company ethics training programs make a difference? An empirical analysis. *Journal of Business Ethics, 11*(9), 719–727.

Diamantopoulos, A. (2011). Incorporating formative measures into covariance-based structural equation models. *MIS Quarterly, 35*(2), 335–358.

Dijkstra, T. K. (2010). Latent variables and indices: Herman Wold's basic design and partial least squares. In V. E. Vinzi, W. W. Chin, J. Henseler, & H. Wang (Eds.), *Handbook of partial least squares: Concepts, methods and applications* (pp. 23–46). New York: Springer.

Edwards, J. R., & Bagozzi, R. (2000). On the nature and direction of relationships between constructs and measures. *Psychological Methods, 5*(2), 155–174.

Efron, B. (1987). Better boostrap confidence intervalls. *Journal of the American Statistical Association, 82*(397), 171–185.

Felo, A. (2001). Ethics programs, board involvement, and potential conflicts of interest in corporate governance. *Journal of Business Ethics, 32*(3), 205–218.

Felo, A. (2007). Board oversight of corporate ethics programs and disclosure transparency. *Accounting and the Public Interest, 7*(1), 1–25.

Fornell, C., & Bookstein, F. L. (1982). Two structural equation models: LISREL and PLS aplied to consumer exit-voice theory. *Journal of Marketing Research, 19*(4), 440–452.

Fornell, C. G., Johnson, M. D., Anderson, E. W., Cha, J., & Byrant, B. E. (1996). The American customer satisfaction index: Nature, purpose, and findings. *Journal of Marketing, 60*, 7–18.

Hair, J. J., Hult, G. M., Ringle, C. M., & Sarstedt, M. (2016). *A primer on partial least squares structual equation modeling (PLS-SEM)*. Thousand Oaks: Sage.

Hair, J. J., Sarstedt, M., Ringle, C. M., & Gudergan, S. P. (2018). *Advanced issues in partial least squares structural equation modeling (PLS-SEM)*. Thousand Oaks: Sage.

Henseler, J., Ringle, C. M., & Sinkovics, R. (2009). The use of partial least squares path modeling. *Advances in International Marketing, 20*, 277–320.

Johansson, J. K., & Yip, G. S. (1994). Exploiting globalization potential: U.S. and Japanese strategies. *Strategic Management Journal, 15*(8), 579–601.

Joseph, J. (2002). Integrating business ethics and compliance programs: A study of ethics officers in leading organizations. *Business and Society Review, 107*(3), 309–347.

Kaptein, M. (2008). Developing a measure of unethical behavior in the workplace: A stakeholder perspective. *Journal of Management, 34*(5), 978–1008.

Kaptein, M. (2009). Ethics programs and ethical culture: A next step in unraveling their multifaceted relationship. *Journal of Business Ethics, 89*(2), 261–281.

Kaptein, M. (2011a). Toward effective codes: Testing the relationship with unethical behavior. *Journal of Business Ethics, 99*(2), 233–251.

Kaptein, M. (2011b). Understanding unethical behavior by unraveling ethical culture. *Human Relations, 64*(6), 843–869.

Kaptein, M. (2015). The effectiveness of ethics programs: The role of scope, composition, and sequence. *Journal of Business Ethics, 132*, 415–431.

Kaptein, M., & Schwartz, M. (2008). The effectiveness of business codes: A critical examination of existing studies and the development of an integrated research model. *Journal of Business Ethics, 77*(2), 111–127.

Kedia, S., Shuqing, L., & Shivaram, R. (2016). Culture of weak compliance and financial reporting risk.

Little, R. J., & Rubin, D. B. (1987). *Statistical analysis with missing data*. New York: Wiley.

MacLean, T., & Behnam, M. (2010). The dangers of decoupling: The relationship between compliance programs, legitimacy perceptions, and institutionalized misconduct. *Academy of Management Journal, 53*(6), 1499–1520.

MacLean, T., Litzky, B. E., & Holderness, D. K. (2014). When organizations don't walk their talk: A cross-level examination of how decoupling formal ethics programs affects organizational members. *Journal of Business Ethics, 128*(2), 351–368.

Majluf, N. S., & Navarrete, C. M. (2011). A two-component compliance and ethics program model: An empirical application to Chilean corporations. *Journal of Business Ethics, 100*(4), 567–579.

McKendall, M., DeMarr, B., & Jones-Rikkers, C. (2002). Ethical compliance programs and corporate illegality: Testing the assumptions of the corporate sentencing guidelines. *Journal of Business Ethics, 37*(4), 367–383.

Menold, N. (2014). *Schriftlich-postalische Befragung*. Mannheim: GESIS—Leibniz-Institut für Sozialwissenschaften.

Nunnally, J. C. (1967). *Psychometric theory*. New York: McGraw-Hill.

Paine, L. S. (1994). Managing for organizational integrity. *Harvard Business Review, 72*(2), 106–117.

Park, H., & Blenkinsopp, J. (2013). The impact of ethics programmes and ethical culture on misconduct in public service organizations. *International Journal of Public Sector Management, 26*(7), 520–533.

Pelletier, K., & Bligh, M. (2006). Rebounding from corruption: Perceptions of ethics program effectiveness in a public sector organization. *Journal of Business Ethics, 67*(4), 359–374.

Ringle, C. M. (2004). Gütemaße für den Partial Least Squares Ansatz zur Bestimmung von Kausalmodellen. Hamburg: K.-W. Hansmann (Hrsg.).

Ringle, C. M., & Sarstedt, M. (2016). Gain more insight from your PLS-SEM results: The importance-performance map analysis. *Industrial Management & Data Systems, 116*(9), 1865–1886.

Ringle, C. M., Wende, S., & Becker, J. M. (2015). SmartPLS 3 [Computer Software]. Retrieved from www.smartpls.com.

Ruiz, P., Martinez, R., Rodrigo, J., & Diaz, C. (2014). Level of coherence among ethics program components and its impact on ethical intent. *Journal of Business Ethics, 128*(4), 725–742.

Schloderer, M., Ringle, M., & Sarstedt, M. (2009). Einführung in varianzbasierte Strukturgleichungsmodellierung: Grundlagen, Modellevaluation und Interaktionseffekte am Beispiel von SmartPLS. In A. Meyer & M. Schwaiger (Eds.), *Theorien und Methoden der Betriebswirtschaft* (pp. 583–611). München: Verlag Franz Vahlen GmbH.

Shih, T. H., & Fan, X. (2008). Comparing response rates from web and mail surveys: A meta-analysis. *Field Methods, 20*(3), 249–271.

Smith-Crowe, K., Tenbrunsel, A. E., Chan-Serafin, S. A., Brief, P., Umphress, E. E., & Joseph, J. (2014), The Ethics "Fix": When Formal Systems Make a Difference. *Journal of Business Ethics, 131*(4), 791–801.

Tobin, G., Breidenbach, W., Klapheke, M., Bentley, F., Pidwell, D., & Simmons, P. (2005). Ethical considerations in the early composite tissue allograft experience: A review of the Louisville ethics program. *Transplantation Proceedings, 37*(2), 1392–1395.

Treviño, L. K., & Weaver, G. (2001). Organizational justice and ethics program "follow-through": Influences on employees' harmful and helpful behavior. *Business Ethics Quarterly, 11*(4), 651–671.

Valentine, S., & Fleischman, G. (2008). Ethics programs, perceived corporate social responsibility and job satisfaction. *Journal of Business Ethics, 77*(2), 159–172.

Warren, D., Gaspar, J., & Laufer, W. (2014). Is formal ethics training merely cosmetic? A study of ethics training and ethical organizational culture. *Business Ethics Quarterly, 24*, 85–117.

Weaver, G. R., & Treviño, L. K. (1999). Compliance and values oriented ethics programs: Influenceson employees' attitudes and behavior. *Business Ethics Quarterly, 9*(2), 315–335.

Weaver, G. R., Treviño, L. K., & Cochran, P. L. (1999a). Corporate ethics programs as control systems: Influences of executive commitment and environmental factors. *Academy of Management Journal, 42*(1), 41–57.

Weaver, G. R., Treviño, L. K., & Cochran, P. L. (1999b). Integrated and decoupled corporate social performance: Management commitments, external pressures, and corporate ethics practices. *Academy of Management Journal, 42*(5), 539–552.

Weber, J. (2006). Implementing an organizational ethics program in an academic environment: The challenges and opportunities for the Duquesne University Schools of Business. *Journal of Business Ethics, 65*(1), 23–42.

Weiber, R., & Mühlhaus, D. (2014). *Strukturgleichungsmodellierung: Eine anwendungsorientierte Einführung in die Kausalanalyse mit Hilfe von AMOS, SmartPLS und SPSS*. Berlin: Springer.

Wright, S. (1921). Correlation and causation. *Journal of Agricultural Research, 20*(7), 557–585.

Wright, S. (1923). The theory of path coefficients: A reply to nils criticism. *Genetics, 8*(3), 239–255.

Wright, S. (1934). The method of path coefficients. *The Annals of Mathematical Statistics, 5*(3), 161–215.

Chapter 6
Why Should PLS-SEM Be Used Rather Than Regression? Evidence from the Capital Structure Perspective

Nur Ainna Ramli, Hengky Latan, and Gilbert V. Nartea

Abstract This study examines capital structure determinants using a simultaneous causal model with interaction effects between manifest and latent variables. Partial Least Squares (PLS) is an approach to Structural Equation Models (SEM) that allows researchers to analyse the relationships simultaneously. It is interesting to compare and contrast this approach in analysing mediation relationships with the regression analysis. In addition to statistical data, logical arguments are presented supported by two case studies from PLS-SEM and regression models. We find that the choice between regression and PLS-SEM matters even with the simplest scenarios per item for constructs. This study's originality is the provision of new comparative analyses of PLS-SEM versus regression analysis in the context of capital structure determinants. The "indirect" and "mediate" macro syntax normal theory of the Sobel test, and the bootstrapping techniques are compared with PLS-SEM. We find that the PLS-SEM analysis provides less contradictory results than regression analysis in terms of detecting mediation effects.

Keywords Capital Structure · Firm Performance · PLS-SEM

N.A. Ramli
Faculty of Economics and Muamalat, Universiti Sains Islam Malaysia, Bandar Baru Nilai, Negeri Sembilan, Malaysia
e-mail: nurainna.ramli@usim.edu.my

H. Latan (✉)
Department of Accounting, STIE Bank BPD Jateng, Semarang, Indonesia

Department of Accounting, Petra Christian University, Surabaya, Indonesia
e-mail: latanhengky@gmail.com

G.V. Nartea
Department of Economics and Finance College of Business and Law, University of Canterbury, Christchurch, New Zealand
e-mail: gerry.nartea@canterbury.ac.nz

N.K. Avkiran, C.M. Ringle (eds.), *Partial Least Squares Structural Equation Modeling*, International Series in Operations Research & Management Science 267, https://doi.org/10.1007/978-3-319-71691-6_6

6.1 Introduction

The academic development of the complexity of hypothesized relationships has been steadily increasing particularly in the social sciences, psychology, accounting and finance. This is because, besides relationships between variables being examined for direct effects, there are many researchers who are more interested in indirect, so-called mediation effects. A mediation effect that implies caution about causality interpretation is defined as a third variable that affects the relationship of the independent with dependent variable. This is depicted in Panel A, Fig. 6.1 as the total direct effect of some cause variable *(X)* on some outcome *(Y)*. Panel B demonstrates the simplest form of mediation. M is the hypothesized mediator between the independent variable X representing firm and country-specific attributes, and dependent variable *Y* representing firm financial performance. Whatever model is used, mediation analysis arises in research when researchers want to see how the impact of variable X on variable Y is affected by another (indirect) variable.

The basic approach for testing the mediation effect in the regression analysis was first presented by Sobel (1982) and Baron and Kenny (1986) (hereafter, BK). Recently studies have tackled this causal issue to improve the basic regression method (Chin 2010; Henseler 2010; Mackinnon et al. 2004, 1995). Iacobucci et al. (2007) claim that there is a better alternative of testing the mediating effects than what is common practice. An alternative to regression for mediation analysis is partial least squares-structural equation models (i.e., PLS-SEM) with endogenous and exogenous variables. Although the literature on the mediation effect in the PLS-SEM model is growing, especially in the social sciences, there is a lack of comprehensive comparisons with traditional regression analysis. Our study offers empirical comparative evidence of these two analytical methods.

Before proceeding to more detail, it is essential to explain the terms mediation and indirect effect as they are commonly used interchangeably. In Fig. 6.1 (Panel A) there is only a direct effect between X and Y, while Panel B is an example of a simple mediation effect which is the direct effect (c') is being mediated by another variable. The indirect effect is mediated by at least one other variable and the total

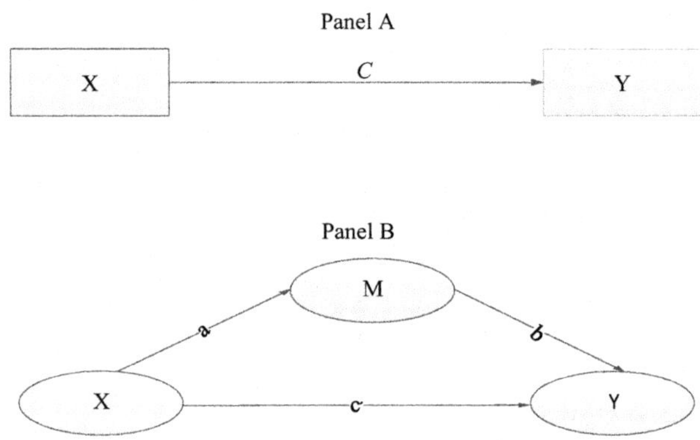

Fig. 6.1 A simple form of mediation effect

effect is the sum of the direct and indirect effects ($c = c' + ab$). The indirect effect ($a \times b$) is estimated statistically from the standardized direct effects that comprise them. For instance, path “a” represents the effect of asset structure (X) on firm leverage (M), path “b” is the effect of firm leverage (M) on performance (Y) and path “c'” is the effect of asset structure on performance through firm leverage. Assuming that the standardized indirect effect ($a \times b$) of asset structure on firm financial performance through firm leverage is the product of 0.0954 (asset structure to firm leverage"X $\rightarrow$ M") and 0.0108 (firm leverage to firm financial performance "M $\rightarrow$ Y" $ab = 1.0303^{-03}$) and that the direct effect (c') between asset structure and firm financial performance is 0.0198 (X $\rightarrow$ Y), then the total effect (c) of asset structure on firm financial performance would be 0.0209 (i.e., 1.0303^{-03}, the indirect effect plus 0.0198, the direct effect).

This study is intended to provide researchers interested in testing for mediation with two different approaches, for studies related to capital structure theory. These approaches are (a) the basic regression and (b) PLS-SEM, which will show the better alternative especially for situations where there are multiple indicators of the *X*, *M* and *Y* variables. The crucial points addressed by this study are: (i) what are the limitations, weaknesses and strengths of each approach and, (ii) which is the better predictive model in capital structure studies. The objective is to develop guidelines for corporate finance researchers applying mediation effect techniques. There is a lack of evidence in the literature that justifies how and why PLS-SEM is the better approach in causal analysis, especially in the capital structure context.

This study is structured as follows. First, we examine capital structure studies that have applied SEM. Second, we use the conceptual model to conduct mediation analysis using regression and PLS-SEM. Third, the regression and PLS-SEM techniques are reviewed. Fourth, we present some prescriptive suggestions for researchers in the area of capital structure. The last section presents a summary of the study.

6.2 SEM Approaches in the Capital Structure Literature

Titman and Wessels (1988) (hereafter, TW) pioneered the use of linear structural equation modelling in examining capital structure determinants (Fig. 6.2). Linear structural equation modelling, known as SEM, involves path analysis, causal models and latent variables. TW estimated eight latent constructs as firm-specific attributes, by measuring 21 indicators, to test their effect on six single latent leverage ratios for a U.S. firm listed in the Annual Compustat Industrial Files. The model is fairly complex; the results tend to have four attributes with a poor correlation and causal relationship with capital structure and the other four with no relationships at all. As a result, they suggest further testing of other indicators of latent variables in future studies. To that end, they strongly mention that *“if stronger linkages between observable indicator variables and the relevant attributes can be developed, then the methods suggested in this paper can be used to test more precisely the extant theories of optimal capital structure” (p. 17)*. This means that finding stronger

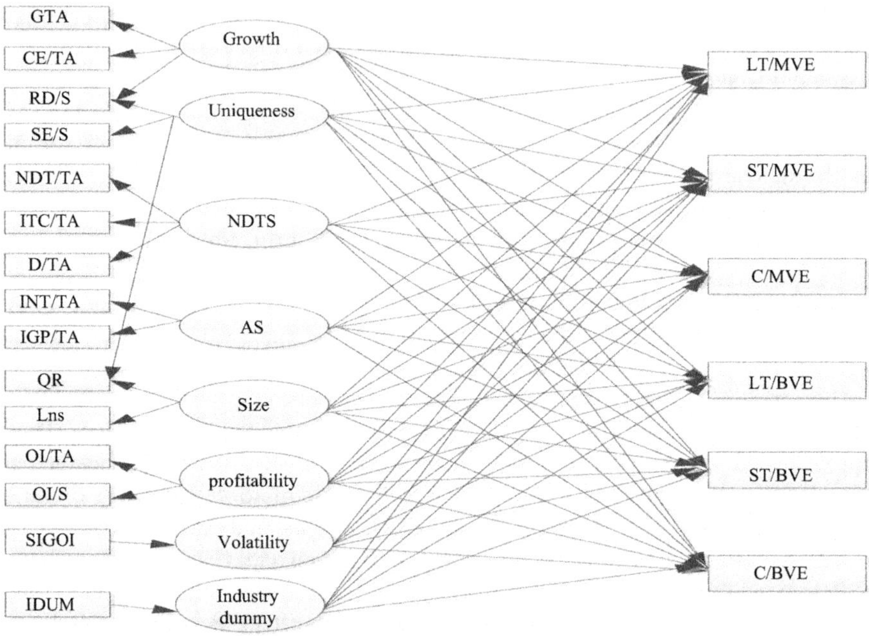

Fig. 6.2 Path diagram implied in Titman and Wessels model (1988)

linkages between indicators and capital structure attributes (LVs) should be developed because there are no unique measures of the attributes represented.

Subsequently, this relatively novel methodology was also applied by Chiarella et al. (1991) (Fig. 6.3). They expanded the technique of linear structural equation modelling in capital structure research by using data from 226 Australian companies from 1977 to 1985. They claimed that the structural model should capture the hypothesized impact of the capital structure attributes on financial leverage. This appears, in general, to agree with TW's study which used the same indicators and technique to examine capital structure determinants. Chiarella et al. (1991) suggested considering more indicators representing each attribute in order to avoid misspecification and obtain more robust results.

Recently the capital structure literature has seen renewed interest in SEM 20 years after it was introduced (Chang et al. 2009; Jairo 2009; Yang et al. 2010). Chang et al. (2009) introduced the so called multiple indicators and multiple causes (MIMIC) model claiming to have a more convincing outcome than TW's study (Fig. 6.4). The poor results in TW's study are mainly caused by problematic model specification that employs too many latent variables for the exogenous and endogenous constructs. By suggesting the MMIC model, the number of LVs may be reduced, for example, from six endogenous LVs to one endogenous LV. Chang et al.'s results show relatively significant relationships with capital structure choice for all seven constructs of firm-specific attributes.

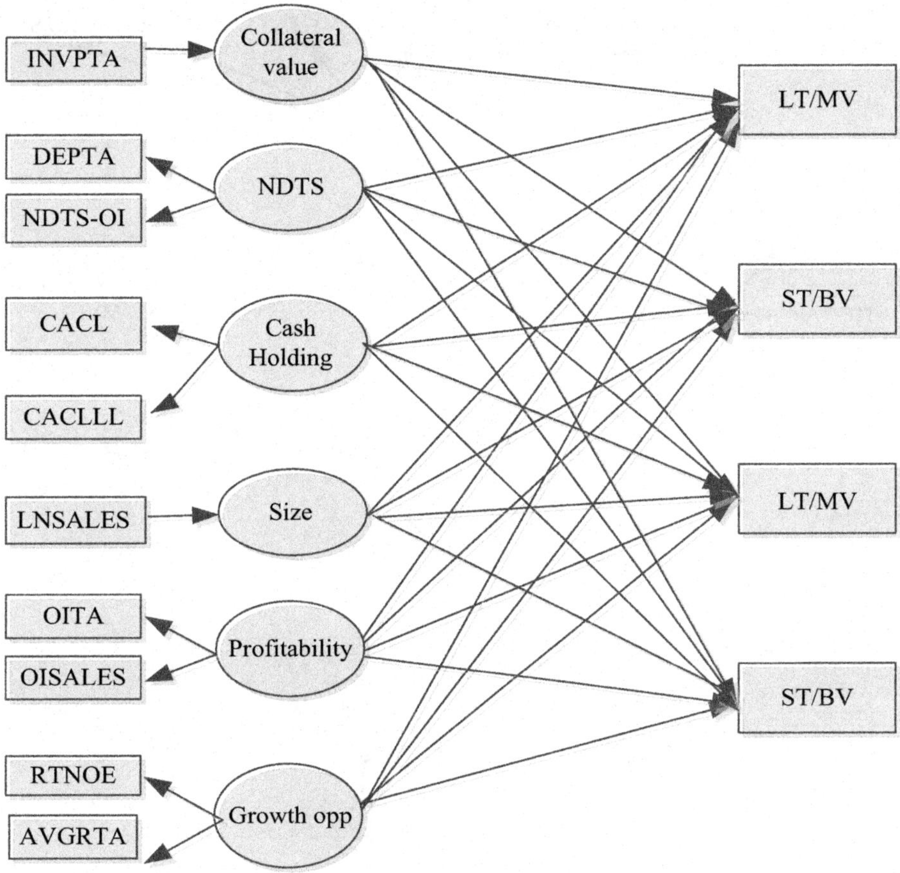

Fig. 6.3 Path diagram implied in the Chiarella et al. (1991) model

Attention to the SEM technique continues to increase in relation to financing decisions in the U.K. market (Fig. 6.5). Jairo (2009) applied factor-analysis for 651 U.K. companies for the 16-years from 1985 to 2000. He observes that most U.K. capital structure research has been conducted using traditional regression estimation such as cross-sectional studies (Bennett and Donnelly 1993; Bevan and Danbolt 2002; Marsh 1982; Ozkan 2001). Traditional regression analysis has been criticised after recent studies involving the estimation model with latent variables, reciprocal causation and interdependence among variables. Jairo (2009) argued that an inappropriate methodology will result in perverse outcomes because of inconsistencies in the choice of firm leverage measures and the proxies for independent variables. Consequently, by attempting new techniques pioneered by TW, most of his findings provide support for the dominant capital structure theories.

Quite recently, the simultaneous model in the SEM technique has been expanded to examine the capital structure determinants in the Taiwan stock markets perspective

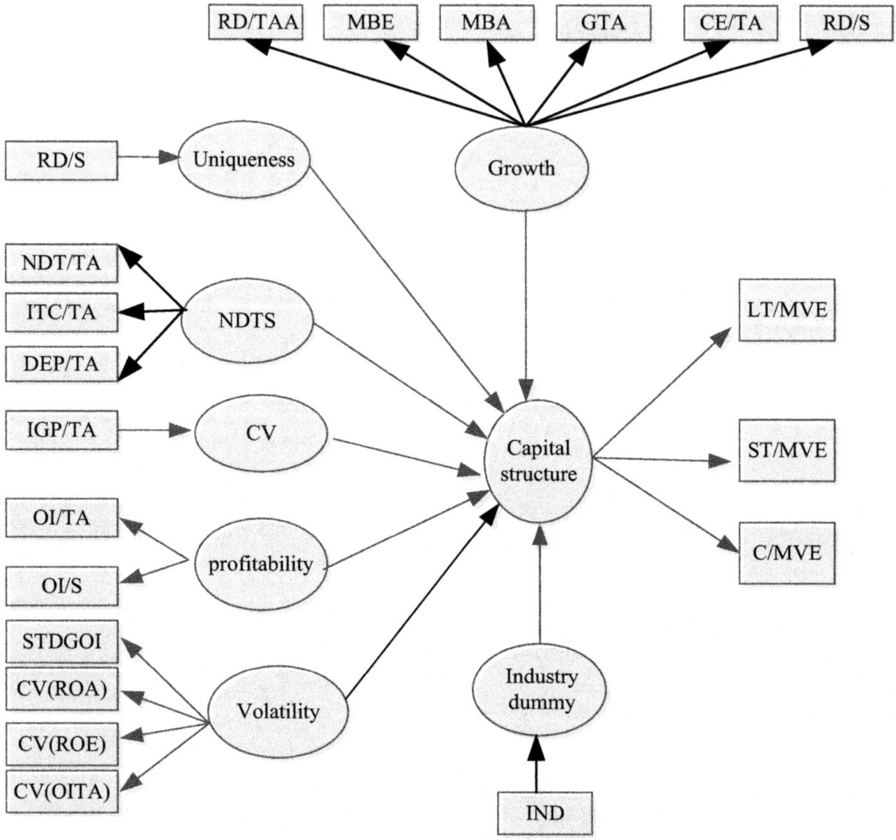

Fig. 6.4 Path diagram implied in Chang et al. (2009) with the MIMIC model

(Yang et al. 2010). Yang et al. (2010) argued that to investigate the effect of firm capital structure on stock returns independently, it is rational to examine the relationships with its common determinants in simultaneous equations (Fig. 6.6). Yang et al. (2010) examined two endogenous variables: capital structure and stock returns and find that the endogenous variables are mutually determined by its factors (i.e., expected growth, liquidity or asset structure). With a sample of 662–702 Taiwanese non-financial companies through 2003–2005, they report that their results are similar to the findings of TW in terms of capital structure determinants and extended it by adding endogenous variable-stock returns in their analysis.

The studies we have examined above indicate that capital structure research commonly used the ordinary least squares (OLS) method with a variety of firm leverage measures as proxies for its dependent variables. However, such a technique could lead to bias in selecting variables as attributes of capital structure based on the statistical goodness-of-fit criteria. Consequently, a bias in economic interpretation

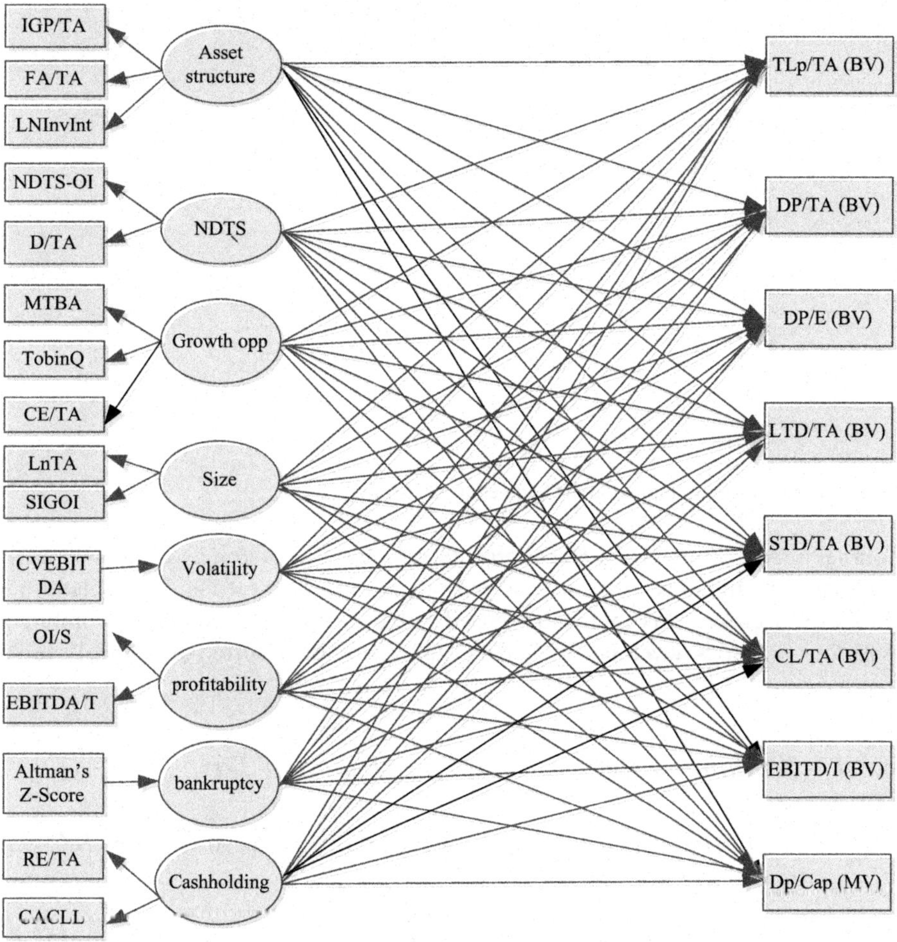

Fig. 6.5 Path diagram implied in Jairo (2009) model

may occur. Most researchers select fewer proxies that have convenient statistics (i.e., have a higher R-squared that shows a high explanatory power).

6.3 Mediation Issues in Regression and PLS-SEM

Testing for mediation effects is routinely done in scientific and social science research published in journals such as the Journal of Consumer Psychology (JCP), Journal of Consumer Research (JCR), Journal of Marketing Research (JMR) and

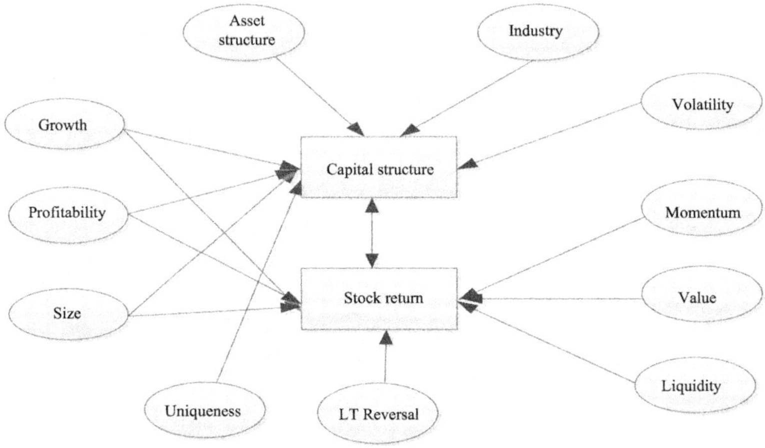

Fig. 6.6 Path diagram implied in the Yang et al. (2010) model

Journal of Marketing (JM). The approach introduced by Baron and Kenny (1986)[1] appears to be the most popular procedure in dealing with mediation effects in the past three decades. This approach is based on basic regression techniques combined with the t-statistics developed by Sobel (1982) (hereafter referred to as Sobel's test).

6.3.1 Traditional Regression in the Mediation Test

The basic regression equations to test for mediation effects as presented by BK need to conform to the following structure:

$$Y = \beta_1 + cX + \varepsilon_1 \tag{6.1}$$

$$M = \beta_2 + aX + \varepsilon_2 \tag{6.2}$$

$$Y = \beta_3 + bM + \varepsilon_3 \tag{6.3}$$

$$Y = \beta_4 + c'X + bM + \varepsilon_4 \tag{6.4}$$

where the betas (β) are the intercept coefficients; *a, b, c* and *c'* are the slope coefficients detailing the linkage among between the three central variables; and the epsilons (ε) are the errors of the models. In the case of simple mediation, evidence from a capital structure perspective is as follows:

[1] See Iacobucci et al. (2007) that report the mediation analysis using Baron and Kenny's (1986) approach in Journal of Consumer Psychology (JCP), Journal of Consumer Research (JCR), Journal of Marketing Research (JMR) and Journal of Marketing (JM).

First equation: the effects measured by the regression coefficient *c (X on Y)*; that is, there is a significant linear relationship between the independent variable (X), which represent from specific attributes and the dependent variable, firm financial performance, denoted as (Y).

Second equation: the effects measured by the regression coefficient *a* (*X on M*); that is, there is a significant linear relationship between the specific attributes (X)-the independent variable and firm leverage (M)-the mediator variable.

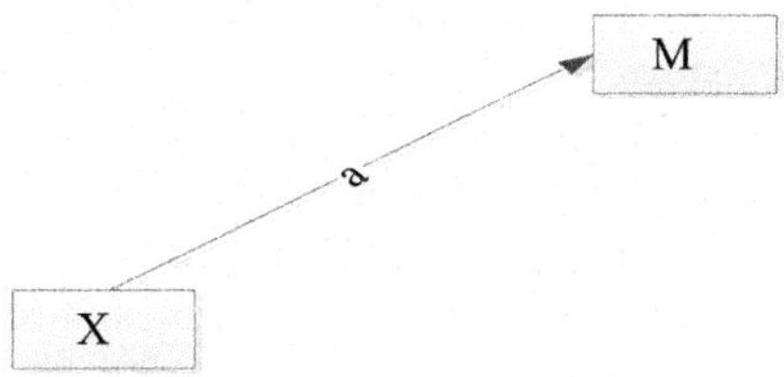

Third equation: the goal is to test the relationship between M that acts as predictor of Y as the criterion variable (estimates for the path *b*).

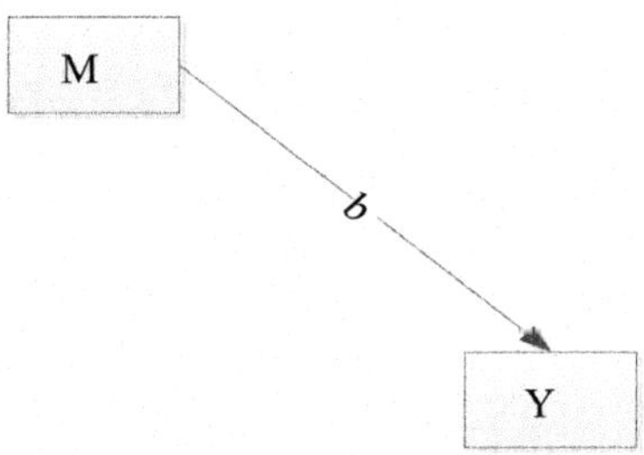

Fourth equation: from the regression coefficient of the term *a, b, c* and also *c' (X on Y controlling M)*; that is, the term c' is the effect of the regression coefficients of specific attributes (X) on the firm financial performance (Y) through firm leverage (M).

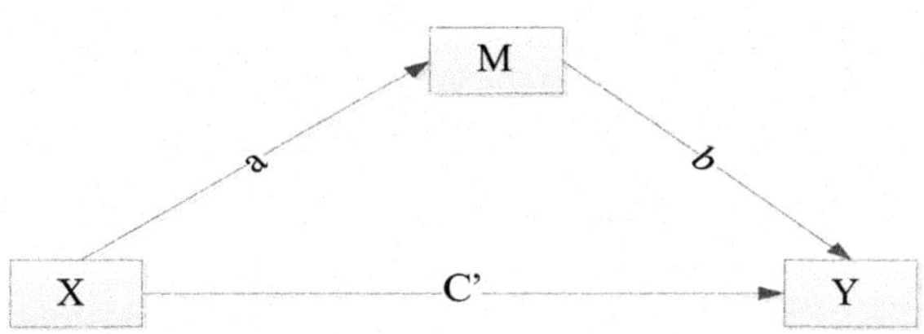

6.3.2 PLS-SEM in Mediation Test

The discussion so far has stated that mediation issues in scientific research have become more complex, particularly involving the third, fourth, or more variables, as those variables might have indirect effects (mediation) on some relationships. From the previous section traditional regression Eqs. (6.1)–(6.4), from BK's approach are applicable to three variables (so-called constructs). That is, each construct is measured by one indicator. In reality, however, most constructs are related to multi-item scales or multiple indicators (they are used interchangeably). For instance, as stated in the corporate leverage literature there is not a single indicator that can represent each construct of interest.

Figure 6.7 illustrates how a focal construct can be measured by multiple indicators (i.e., X is the exogenous variable with multiple items of $X_1 \ldots X_3$; M is the mediator variable with multiple items of $M_1 \ldots M_3$; Y is the dependent variable with multiple items of $Y_1 \ldots Y_3$). This figure demonstrates that two, three or more items can explain the attributes of interest—X, M or Y, and the number of indicators is not necessarily equal for each construct. Following the example from Iacobucci et al. (2007) and Hayes (2013), let us assume that there are three predictor variables, X_1, X_2 and X_3, a single mediator M, and a dependent variable Y. The structure set in Eqs. (6.1)–(6.4) can now be expressed as follows:

$$Y = \beta_5 + c_1 X_1 + c_2 X_2 + c_3 X_3 + \varepsilon_5 \tag{6.5}$$

$$M = \beta_6 + a_1 X_1 + a_2 X_2 + a_3 X_3 + \varepsilon_6 \tag{6.6}$$

$$Y = \beta_7 + bM + \varepsilon_7 \tag{6.7}$$

$$Y = \beta_8 + c'_1 X_1 + c'_2 X_2 + c'_3 X_3 + bM + \varepsilon_8 \tag{6.8}$$

Like any regression, this basic approach to regression equations would raise a few concerns from a statistical viewpoint. The first issue is which coefficients should be

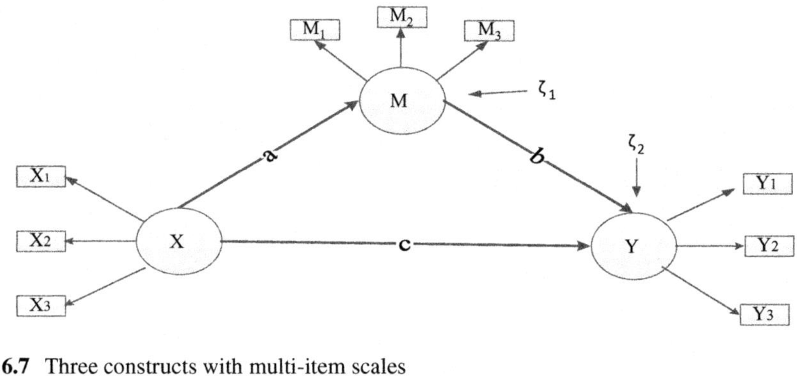

Fig. 6.7 Three constructs with multi-item scales

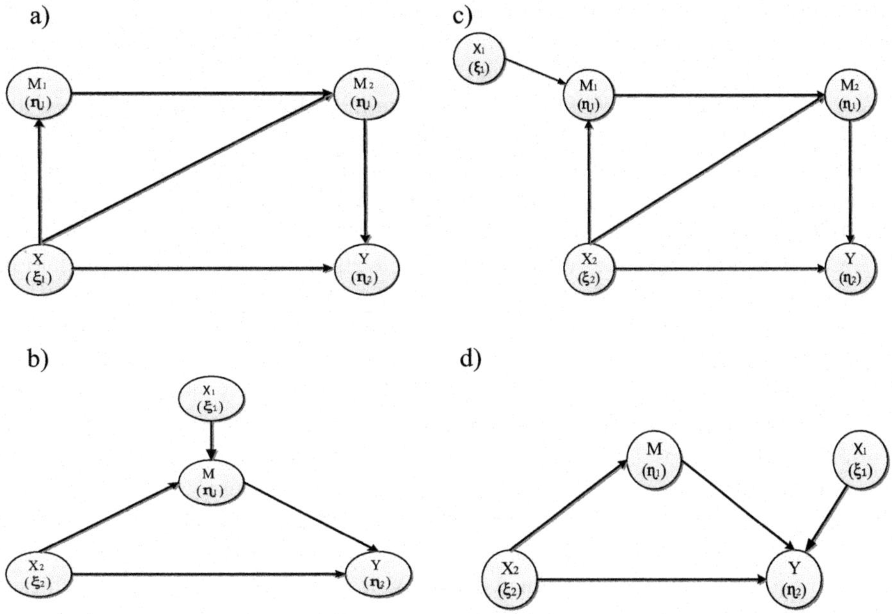

Fig. 6.8 A larger nomological linkages to test mediation effects

associated with the mediation effect and how are they related. The second issue is the possibility of having a multi-collinearity problem among the item-scales (i.e., $X_1..X_n$). This simple model can be even more complex with the introduction of multiple constructs such as having two or more mediator variables (i.e., $M_1, M_2 \ldots M_n$) as in Fig. 6.8. This larger nomological linkage encourages one to investigate a broader network to test mediation effects, particularly for multivariate analysis (i.e., X → Y, or, M → Y, or, Y → M, or, X → M, or, M → X). However, this issue will not be highlighted any more deeply in this study because it is more important to gain an understanding of the basic elements as in a simple mediation model. Figure 6.7 can be extended to explain an example of the complex model as in Fig. 6.8 for mediation effects.

In the first section, it was mentioned that Titman and Wessels (1988) study used an innovative approach for a cause-and effect model in capital structure that is designed for multivariate analysis. PLS-SEM is a well-known technique for multivariate analysis that can extend the simpler univariate prediction. The use of PLS-SEM in solving a multivariate system such as in Fig. 6.8 has increased greatly. Particularly, PLS-SEM is renowned for structuring mediation effects with multiple indicators. This is because, Baron and Kenny (1986) argued that, *"the common approach to unreliability is to have multiple operations or indicators of the construct"*. One can use the multiple indicator approach and estimate mediation paths by the latent-variable structural modelling method. BK argued that there are another three key advantages of using the SEM operation. First, SEM is designed for non-experimental data (i.e., a correlation study). Second, most possible indicators can be tested and not be omitted. Third, it minimizes the complications of measurement error and correlated measurement error in the model.

The traditional regression approach does not address the difference between the direct effect (*c*') and indirect effect (*ab*) compared with the mediation effect of the SEM approach (Hayes 2009; Preacher and Hayes 2004; Vanderweele 2015).

Recent studies have proposed few approaches to test the mediation hypothesis (Cepeda et al. 2017; Nitzl et al. 2016; Preacher and Hayes 2004, 2008; Zhao et al. 2010). In essence, the mediation test focuses on the product term *ab*, the difference between the total and direct effects. The first well-known approach in mediation t-statistics is the Sobel test (Sobel 1986, 1982), also called the *product-of-coefficients approach*. In this approach, the t-statistic is computed with the assumption of a normal distribution and a significant value supports the hypothesis of mediation. However, methodologists have taken issue with the validity of the t-statistic using the normal distribution, since the normal distribution is uncommon except for large samples. Thus, a non-parametric resampling approach, such as *bootstrapping*, has become an additional method in testing for mediation (Hair et al. 2017; Jose 2013; Preacher and Hayes 2004, 2008; Zhao et al. 2010). The assumptions in this approach do not impose normality on the sample distribution. The bootstrapping technique will be discussed shortly. In extensive studies of set simulation, Mackinnon et al. (2002, 2004) strongly recommended the use of bootstrapping over the Sobel test or causal step approach[2].

Therefore, to address the mediation issue in traditional regression we apply both approaches to obtain more robust results and to avoid erroneous classifications. That is, we test the mediation effect by using the Sobel test[3] and the 95% bias corrected bootstrap confidence interval (hereafter, 95% BC bootstrap CI) method. While, in PLS-SEM we use 95% BCa bootstrap CI method.

6.3.2.1 Sobel Test

The Sobel test formula was used to obtain a robust result[4]. The Sobel test uses t-statistics to test for the presence of mediation effects. For example; if the effect of specific attributes on firm financial performance decreases to zero with the inclusion of the mediator variable, firm leverage, then perfect mediation is said to have occurred. However, there is a possibility of partial mediation in which the effects of

[2] Mackinnon et al. (2002, 2004) included these methods to examine the performance on their type error rates and power.

[3] The mediation bootstrap t-statistic is computed from macro syntax in SPSS, the bootstrap test provides summary results of percentile-based and bias-correlated bootstrap confidence intervals. From a statistics point of view, interval estimates of a popular parameter estimate the reliability. The underlying confidence interval contains the two-sided confidence limits from a confidence interval in which their one-sided counterparts are denoted as lower or upper confidence bounds.

[4] The outcome coefficient and standard error estimated by bootstrapping is used in computing the Sobel statistical value.

specific attributes on firm financial performance decrease to a nontrivial amount, i.e., the value of the linear regression coefficient of X on Y is not equivalent to zero. The Sobel test statistic is;

$$Sobel = \frac{a \times b}{\sqrt{b^2 s_a^2 + a^2 s_b^2}}$$

where "a" is the slope coefficient for the association between independent (X) and mediator variable (M) and "s_a" is the standard error of a, "b" is the slope coefficient for the association between mediator (M) and dependent (Y) variable (when the independent variable is also a predictor of the dependent variable) and "s_b" is the standard error of b. "a" and "s_a"and "b" and "s_b" are taken from Eq. (6.4) (see Sect. 6.3.1). The Sobel t-stat can be calculated on the available website (Preacher 2013 www.danielsoper.com/statcalc3/calc.aspx?id=31 or, quantpsy.org/sobel/sobel.htm). The assumption in this approach is that if one of the first two equations is not significant (i.e. either a or b), then there is no mediation effect. Iacobucci et al. (2007) claimed that 67% of studies in consumer and psychological research that test the mediation effect properly follow rules (1) to (4) However, they found that 89.7% have not reported the t-statistics.

6.3.2.2 Bootstrapping

The procedure outlined by BK using the Sobel t-statistics in testing for mediation effects is already well-known in the social science literature. However, recent studies by Preacher and Hayes (2004, 2008) have shown renewed interest in mediation issues with Cepeda et al. (2017) and Hair et al. (2017) proposing an alternative approach to assessing mediation relationships. The alternative approach is to bootstrap the sampling distribution of paths "a" and "b" and derive empirically the confidence interval from the bootstrap sampling distribution. This approach is a nonparametric resampling approach with no assumptions regarding the sampling distribution (Hair et al. 2013, 2012; Preacher and Hayes 2008). The bootstrapping procedure is usually done to deal with non-normality in the sampling distribution. It can be reasonably applied with confidence for small or large samples. Bootstrapping is done by sampling with replacement from the original data set and then the direct effect of *ab* is calculated for each sample set (Preacher and Hayes 2004). The bootstrap samples are set for 1000 or 5000. The mean path ab is computed over the 1000 or 5000 samples and the estimated standard error is taken from the standard deviation of the 1000 or 5000 *ab* estimates. To set the significance level of the mediation effects, a 95% confidence interval is derived from the 1000 sample estimates of path *ab*. For example, say the point of the indirect effect estimates generated from the raw data lie between 0.0458 and 0.8123 with a 95% confidence interval (CI). Then, because zero is not included in the 95% confidence interval, one can conclude that that the mediation or indirect effect is indeed significantly different from zero at

$p<0.05$ (two tails). Therefore, in testing the mediation effect, the researcher is advised to consider a procedure that works from the bootstrapping sampling distribution proposed by Preacher and Hayes (2004, 2008). The procedure can be used for all models whether simple or multiple and whether we have large or small data sets.

6.4 Comparison of Results from Traditional Regression and PLS-SEM Techniques

To further investigate the benefits of implementing the SEM versus regression analysis, two simulated case studies on capital structure are conducted. The main focus of this study is to compare a simultaneous approach over a piece-meal method.

The sample size for these two case studies is 7819 firm-year observations from two subsample countries (Malaysia and Indonesia) for the 20-year period from 1991 to 2010. Keeping the main three latent variables of *X, M* and *Y*, these case studies were conducted by selecting the first latent variable '*X*', firm size, as a firm-specific factor and economic growth as a country-specific factor. Firm size is measured by the logarithm of total assets and the logarithm of total sales, and, economic growth is measured by the annual percentage growth rate of the gross domestic product (GDP) and gross domestic investment (GDI), i.e., the addition of the fixed assets of the economy plus the changes in the inventories. The second latent variable, '*M*', is firm leverage with seven factors: total debt ratio (TADR), debt to total capital of book (TDTC-BV) and market value (TDTC-MV), long term debt to capital of book (LTDTC-BV) and market value (LTDTC-MV) and short term debt to capital of book (STDTC-BV) and market value (STDTC-MV). The third latent variable, '*Y*', is firm financial performance calculated from the return on equity (ROE), return on assets (ROA) and return on investment (ROIC).

The two simulated case studies compare the ability of PLS-SEM and regression analysis in detecting mediation effects. In the first case study, each construct (LV) is analysed separately using regression and PLS-SEM techniques. The second case study conducts the analysis using multiple indicators for each construct. The case studies are as follows.

6.4.1 *Study 1*

The analytical approach comparing the PLS-SEM and regression techniques investigates the strength of specific items for the direct and indirect effect of firm leverage between the capital structure determinants and firm financial performance with the simplest data scenario (i.e., one X, one Y and one M). This scenario analyses each item's attributes in the regression and PLS-SEM techniques one at a time.

In this capital structure study there are seven possible measurements of firm leverage (M)[5]. MacKinnon et al. (2004, 2002) recommended the use of bootstrapping because of its ability in controlling the Type I error. Type I error, when comparing means, would be rejecting the true null hypothesis. For example, if the null is that the means are the same, a Type I error is committed if we conclude that the means are different when they are not. However, for resampled datasets (bootstrapped) one would strongly suggest conducting the test with small samples and not for large data sets (Preacher and Hayes 2004, 2008 for an overview).

6.4.1.1 Results of Study 1

Traditional Regression

The causal step popularized by Baron and Kenny (1986, 1982) using OLS regression is extended to meet several criteria. The first to third criteria (Eqs. (6.1)–(6.3)) must hold in order to satisfy the mediation product. However, if first to third criteria (Eqs. (6.1)–(6.3)) hold and forth criteria (Eq. (6.4)) which is c' is not significantly different from zero, the effect will be concluded to be as a perfect or complete mediation. Judd and Kenny (1981) argued that the difference of the coefficient to control for the debt level can also exist to signpost mediation effects[6]. These criteria have received attention from some authors who argue that an insignificant total effect *c* does not necessarily mean that mediation occurred (see Judd and Kenny 1981; Mackinnon 2000; Mackinnon et al. 2002, 2004; Shrout and Bolger 2002). This point of argument will be explained in more detail shortly. Table 6.1, which is taken from the full macro output in Table 6.8 of Appendix 6.1 (see for the whole view, the total effect in path *c* of IV on DV) shows non-significant estimates for both measures of firm size (i.e., total assets and total sales), yet it is still possible for mediation to occur[7].

Table 6.1 shows the macro syntax matrix regression about the process implications of mediation effects. First, Table 6.1 confirms that it is not a necessary requirement for the total path *c* to be significant for mediation to occur. For instance, from Table 6.8, the total effect of IV on DV (path *c*) shows that firm size measured by total assets (path $c = t = -1.7689$) and total sales (path $c = t = -1.4268$), are not significant. The effect even becomes smaller (and still insignificant) after controlling for the ratio of firm leverage, i.e., path $c' = t = -0.5823$and $t = -0.3307$

[5]The regression result will show of total debt ratio (TADR), followed by debt to total capital of book (TDTC-BV) and market value (TDTC-MV), long term debt to capital of book (LTDTC-BV) and market value (LTDTC-MV) and short term debt to capital of book (STDTC-BV) and market value (STDTC-MV), respectively.

[6]*c*', that is "$c'=c-ab$" defines the difference between the total effect (*path c*) and indirect effect of X on Y via M (path *ab*).

[7]This can be seen in both summary results from the bootstrap technique with percentile-based, bias-correlated (BC) bootstrap confidence intervals (CI) or from the normal theory of causal effects in Table 6.8 of Appendix 6.1

Table 6.1 Macro syntax matrix regression procedures[a] (Preacher and Hayes (2008) defined the SPSS macro syntax matrix as "*this macro estimates the path coefficients in a multiple mediator model and generates bootstrap confidence intervals (percentile, bias-corrected, and bias-corrected and accelerated) for total and specific indirect effects of X on Y through a one or more mediator variable(s) M. This macro is far superior to SOBEL, as it allows for more than one mediator and adjusts all paths for the potential influence of covariates not proposed to be mediators in the model*". A SPSS macro function is a "mini program" within the syntax of SPSS. Macros are quite complex to perform by issuing commands one step at a time. The SPSS macro language is also quite flexible. Macros able to perform a whole series of analyses, adjustments for the multiple output statistics, perform multiple charts with many combinations of variables and numbers (see http://wise.cgu.edu/spssmacro/index.asp#what or http://www.quantpsy.org/medn.htm).)

A. Total effect of IV on DV (c path)

X to Y	Coef. B	Std.error	T-value	P-value
Firm size (log total assets)	−2.7306	1.5437	−1.7689	0.0770
Firm size (log total sales)	−2.0990	1.4712	−1.4268	0.1537
Economic growth (GDP)	4.0510	0.6231	6.5018	0.0000
Economic growth (GDI)	0.7105	0.3801	1.8691	0.0616

B. Direct effect of IV on DV controlling for mediator (c' path)

Model	Mediator		Coef. B	Std.error	T-value	P-value
A	TADR	Firm size (log total assets)	−0.9194	1.5789	−0.5823	0.5604
		Firm size (log total sales)	−0.4956	1.4988	−0.3307	0.7409
		Economic growth (GDP)	3.8027	0.6240	6.0938	0.0000
		Economic growth (GDI)	0.7146	0.3794	1.8833	0.0597

C. Normal theory tests for mediation effects

Path coefficient	Sobeltest[a]	Goodman test[b]	Percentage of the total effect that is mediated	Ratio of the indirect to the direct effect
Sizeta → TADR → ROE	−5.0957	−5.0895	66.3305	1.9700
Sizesales → TADR → ROE	−5.1318	−5.1247	76.3888	3.2353
GDP → TADR → ROE	4.0932	4.0660	6.1302	0.0653
GDI → TADR → ROE	−0.1708	−0.1681	−0.5716	−0.0057

D. Model summary for DV Model- bootstrap technique

			Bias corrected (BCa) a confidence intervals (CI)		Bootstrap results for mediation/indirect effects
Model	Mediator	DV=ROE	Lower	Upper	
A	TADR	Firm size (log total assets)	−5.2690	−0.5492	Support for mediation
		Firm size (log total sales)	−4.1018	−0.5123	Support for mediation
		Economic growth (GDP)	0.0684	0.6876	Support for mediation
		Economic growth (GDI)	−0.0730	0.0328	No support for mediation

Note- [a]*is the Sobel Test of t-value* = $a*b/SQRT\ (b^2*s_a^2 + a^2*s_b^2)$ *and* [b]*is the Goodman Test of t-value* = $a*b/SQRT\ (b^2*s_a^2 + a^2*s_b^2 - s_a^2*s_b^2)$

Note- 1000 bootstrap samples, BC, bias corrected with 95% level of confident for intervals

[a]The full macro syntax matrix regression procedures can be seen in Appendix 6.1 of Study 1. This Table reveals only a small part of the output, i.e., only one item measure of firm leverage, TADR (M), with the associated specific variables (X) and firm financial performance (ROE = Y)

(see panel B of Table 6.1). However both the Sobel Test (−5.0957) and the Goodman Test (−5.0895)[8] appear to be significant, i.e., firm leverage might mediate the relationship between firm size and firm financial performance[9]. As mentioned earlier, an alternative procedure for testing the mediation effects is through the bootstrap technique. Also, the bootstrap technique (panel D) produces the same outcome as the normal theory test for a mediation effect, i.e., the Sobel and Goodman Tests (panel C) show the existence of mediation (both panels are from Table 6.1). The bootstrap requires 1000 resamples for a 95% level of confidence interval. For example, the output for the 95% bias corrected bootstrap confidence interval (hereafter, 95% BCa bootstrap CI reported in panel d of Table 6.1) is as follows: the result of firm size (log total assets) on performance (ROE) for the *lower* limit of the bootstrap *confidence interval* (**BootLLCI**) is −5.2690 and the *upper* limit of the bootstrap *confidence interval* (**BootULCI**) is −0.5492. Since the values "BootLLCI" and "BootULCI" do not straddle "0", the outcomes show that mediation occurred (Preacher and Hayes 2008).

Second, the single-item scales deliver outcomes for each item individually that would clarify which items or indicators might explain particular relationships. Single item scales often too easily provide empirical support for almost any relationships. However, it is not usual for the overall understanding of the theory of capital structure determinants to deal with several items to represent each determinant. For instance, it is common from the capital structure literature to use one dependent variable (formula ratio, i.e, TADR) for the term "firm leverage" or "capital structure". Even if the studies use more than one dependent variable, they are more likely to analyse dependent variables one by one. In this scenario, the term "firm leverage" maybe misleading. For example, firm leverage might have seven possible measurements and different firm leverage measures (i.e., total debt ratio (TDR) or long term debt to capital (LTDTC)). Total debt ratio (TDR) may not be a significant mediator but long term debt to capital (LTDTC) may be a significant mediator. If there is a contradiction in the results among the different measures of firm leverage, how can we conclude whether or notthe term 'firm leverage" (M) is a significant mediator variable between the relationship of X and Y? So, total debt ratio (TDR) or long term debt to capital (LTDTC) cannot simply be used to refer to the term "firm leverage" or "capital structure". Therefore, for this scenario, instead of using the general word "firm leverage", it is more useful to refer to particular measures of firm leverage such as long term debt or short term debt in defining relationships.

Third, the output from testing the mediation effects (see Appendix 6.1 in Table 6.8) from the normal theory (i.e., the Sobel or Goodman test) and the bootstrapping approach are inconsistent in their t-statistic interpretation. For instance, the bootstrap output shows that the mediation effect for long and short term debt

[8] The Goodman Test is a robustness test that is also commonly used to test mediation effects, but it is not as popular as the Sobel Test.

[9] The test statistics from the Sobel and Goodman Tests were generated from the macro documented at http://www.comm.ohio-state.edu/ahayes/sobel.htm.

total capital on market value (LTDTC-MV and STDTC-MV) is different from zero with the 95% BCa bootstrap CI, but the Sobel test indicates that it is not. This inconsistency is quite problematic. Nevertheless, output directions of the "*a*" and "*b*" paths in Appendix 6.1 are consistent with the interpretation of various capital structure theories (i.e., TOT and POT) on such a relationship.

PLS-SEM

The extensive multivariate tool, PLS-SEM, was used to compare the performance of a single simultaneous structural model for each firm leverage measure with a total of 28 PLS-SEM models[10] The model is similar to regression, i.e., the simplest data scenario with one to one variables[11]. The difference from regression analysis is that each mediator parameter estimate for firm leverage is simultaneously measuring exogenous variables-, i.e., firm size and economic growth concurrently with the endogenous variable, -firm financial performance.

In comparing PLS-SEM versus regression analysis, we find the following. First, the standard errors of PLS-SEM coefficients are smaller than those from regression analysis (Fig. 6.9). For example, with the same models[12], the standard error of the regression equation between X and Y (path *c*) is 1.5437 (see Table 6.1) compared with 0.021 for PLS-SEM (Fig. 6.9). Figure 6.10 shows the difference in standard errors from the regression analysis (Appendix 6.1 in Table 6.8) versus PLS-SEM (Appendix 6.1 in Fig. 6.14) for path "*c*"($X \rightarrow Y$) for each firm leverage measure. The PLS-SEM results are preferable because of their smaller standard error. The reason might be because regression analysis involves several individual regressions whereas PLS-SEM is a simultaneous approach. Iacobucci et al. (2007, p. 146) stated that *"fitting components of models simultaneously is always statistically superior to doing so in a piece-meal fashion…"*, and they argue that a simultaneous approach produces more consistent estimates that are closer to the true values. Thus, the powerful simultaneous approach might have been able to reduce standard errors.

Second, by using the 95% BCa bootstrap CI, we find that the results from traditional regression analysis and PLS-SEM show inconsistent significance estimates. Take an example in Table 6.8 (Appendix 6.1) for firm size (sales), the bootstrap for a mediation effect for long term debt total capital on market value (LTDTC-MV) is 95% BCa CI of −0.297 (bootLLCI) to −0.1337(bootULCI) is significant (since the values do not straddle 0 in the confidence intervals, the conclusion is that there is a mediation) but the PLS-SEM in Appendix 6.1 is not (95% BCa CI = 0.092). Thus, the results of the mediation effect from both PLS-SEM and regression could cause more controversy and difficulty. This is important to highlight that PLS-SEM method is considering the significant relationship of $M \rightarrow Y$(path *b*) through the inclusion of a third variable (X). This supports the mediation criteria arguments

[10] See Appendix 6.1 in Fig. 6.14 for complete results.

[11] Simplest data (i.e., one item for variable X, one item for variable M, one item for variable Y).

[12] X = firm size (total assets), M = TADR, Y = ROE.

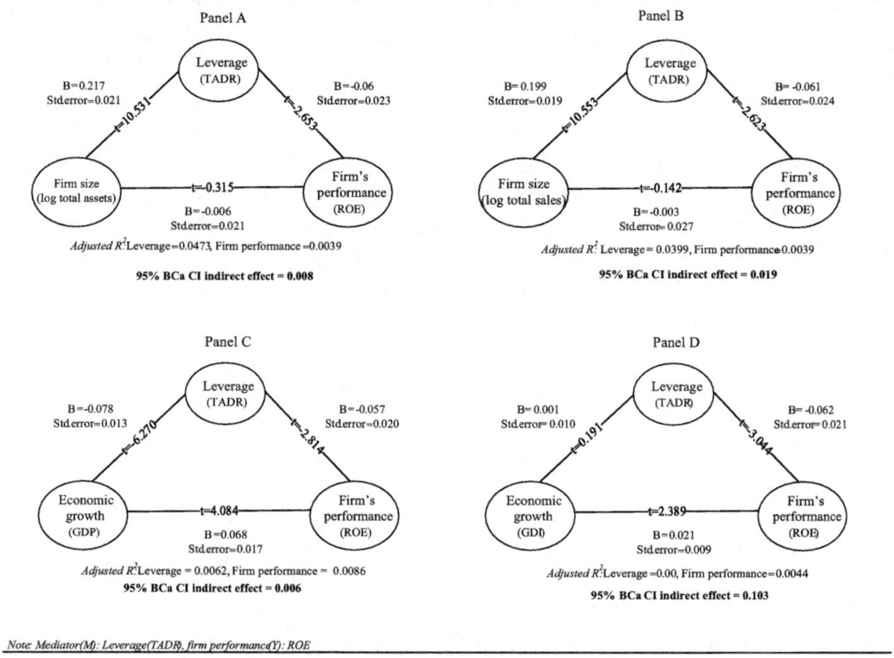

Fig. 6.9 Single simultaneous equation model in PLS-SEM

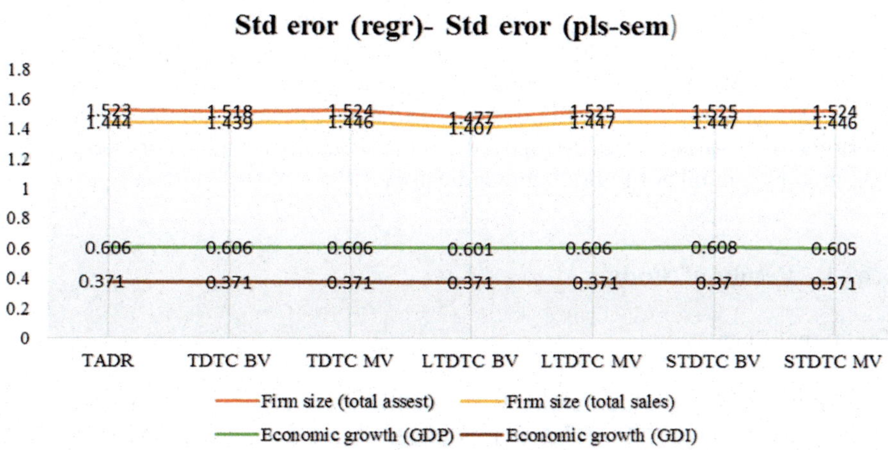

Fig. 6.10 A comparison of regressions and PLS-SEM standard errors

from prior studies (see Baron and Kenny 1986; Iacobucci and Duhachek 2003; Mackinnon et al. 1995). For example, if the path coefficients from the PLS-SEM model for $X \rightarrow Y$and $X \rightarrow M$were significant and $M \rightarrow Y$were not, then the bootstrap value will not be significant. Note too, with the single simultaneous production of a predictor (X), mediator (M) and criterion (Y) variables that fit in the same model, all the standard errors, parameter estimates and t-values are evaluated simultaneously in the model.

In sum, the choice between regression and PLS-SEM matters even with the simplest scenarios per item for each of the three constructs. PLS-SEM is superior to regression analysis because, one is more likely to spot whether there is mediation by simply looking at the model diagram (see Fig. 6.9 for the examples) whereas in regression analysis we need to set up four models separately[13]. Case Study 2 below delves more on this issue. In addition, this study supports Iacobucci et al.'s (2007) argument that PLS-SEM estimates tend to be true and close to the known population structural characteristics and finally be more convincing and reliable with simultaneous statistical estimations.

6.4.2 *Study 2*

In this section, we compare the ability of PLS-SEM and regression analysis to detect mediation effects using multiple indicators per construct (latent variables). A scale performance for multiple items per construct should also be verified (i.e., two or more X, two or more M, two or more Y). According to statistical theory, multiple item scales produce stronger results than those with single items. This case study continues the design built in Study 1, i.e., the same sample population estimates and variables, and extends the examination to multiple items measuring a construct. The only difference from the Study 1 is that we measure firm financial performance with multiple indicators namely, ROE, ROA and ROIC. The reason for adding the items for firm financial performance is to align with the concept of multivariate tools of PLS-SEM that allows the design of latent variables. A latent variable (also called construct) is a variable that is inferred from the measured indicators/items or is measured by a number of observable indicators.

6.4.2.1 Results of Study

Traditional Regression

The outputs of the mediation effects for multi-item scales are obtained from the output macro syntax SPSS. This step is identical to case Study 1, except that the regression program is run less number of times because the regression is run in multiple equations. Because errors-in-variable (EIV) and multi-collinearity are present, a number of separate regressions were conducted; this will be discussed shortly. The SPSS macro syntax version for Study 2 was obtained from the "mediate" macro- http://www.afhayes.com/spss-sas-and-mplus-macros-and-code.html.

[13] When the model diagram shows that at least paths "a" and "b" are significant, there is strong support that the mediation exists. The significance of the mediation can be confirmed with the t-statistics test.

In this case study, we confirm that the multi-item scales result in multi-collinearity problems. Multi-collinearity is a problem when the correlation between two independent variables is 0.9 or above (Latan 2015). Multi-collinearity can be assessed through the product of tolerance or VIF for each of the independent variables in the multiple regressions. According to Hair et al. (2010) and Lin (2008), the rule of thumb is that a VIF more than 10 and have less than 0.10 of tolerance (VIF = 1/TOL) this indicates that the independent variables have a high degree of co-llinearity or multi-collinearity. Caution in interpreting results is highly suggested if the tolerance values are in between 0.20 and 0.10. Also, if the tolerance values are less than 0.10 and the VIF values are more than 10, this indicates a serious multi-collinearity problem, and it is recommended that one reconsider the choice of independent variables Hair et al. (2010).

Based on the multi-collinearity statistics matrix in the VIF column shown in Table 6.2, the predictors of firm size, measured either by log total assets and total sales, the determinants of capital structure (X), are highly correlated with firm financial performance (Y)[14]. For example, in Table 6.2 the result of Part 1 reveals that the VIF for firm size (log total assets) is 19.920 and that for firm size (log total sales) is 19.436. These conditions are indicative of a multi-collinearity problem. Although some researchers claim that this problem is not so crucial, erroneous empirical analysis might result from the data. Based on Neter et al. (1990), the linear regression assumption might not be violated but other difficulties might arise. The difficulties proposed by Lin (2008, p. 417) are as follows: (i) unreasonably large variances of the parameter estimates; (ii) this might produce insignificant parameter estimates; and (iii) the sign coefficient parameter estimates may be different from what is anticipated. This statistical problem would result in misleading data analysis and might deceive researchers in their analysis if they do not report it.

Therefore, to be acceptable from the practical and statistical points of view and being free of multi-collinearity, the robustness test for both measures of firm size (i.e., log total assets and sales) is analysed individually. Based on multicollinearity diagnosis, the regressions are done in three parts. In Part 1 the regressions are done using all predictors, hence embracing a possible multi-collinearity problem. In the next two regressions, we mitigate the problem of multicollinearity by dropping log total sales and in Part 2 and dropping log total assets in Part 3. The results for all three regressions seem to be basically the same. Although the outputs are alike, the arguments about the multi-collinearity problem discussed earlier, especially on significance of the parameter estimates, are evident. For example in Table 6.2, the result of path "*a*" in Part 1: firm size measured by total assets and total sales (predictor variable-X) and TDTC-MV (mediator-M) is not significant (t = 1.6447 and t = 0.269, respectively) but does it is in Part 2 (t = 8.0772) and 3 (t = 7.8948). Lin (2008, p. 417) argues that "*in the process of fitting regression model, when one independent variable is nearly combination of other independent variables, there will affect parameter estimates*". Another example is between the X of economic growth (GDI) and M of STDTC-MV. Part 1 shows that the relationship does not seem to be

[14] The collinearity statistics matrix (VIF) also estimates other relationships such as capital structure determinants; they have consistent results.

Table 6.2 Macro syntax matrix for multiple regression procedures

	Part 1: With multi-collinearity					
	IV to Mediators (*a* paths)					
Mediator	X to M	Coef. B	Std.error	t-value	p-value	Adj. R^2
TDTC MV	Constant	0.7486	0.0205	36.5840	0.0000	0.0088
	Firm size (log total assets)	0.0169	0.0102	1.6447	0.1001	
	Firm size (log total sales)	0.0026	0.0097	0.2691	0.7879	
	Economic growth (GDP)	−0.0022	0.0011	−2.0513	0.0403	
	Economic growth (GDI)	−0.0001	0.0007	−0.1370	0.8910	
STDTC MV	Constant	0.7452	0.0222	33.5963	0.0000	0.0102
	Firm size (log total assets)	−0.0207	0.0111	−1.8624	0.0626	
	Firm size (log total sales)	0.0397	0.0105	3.7739	0.0002	
	Economic growth (GDP)	−0.0026	0.0012	−2.2318	0.0257	
	Economic growth (GDI)	−0.0009	0.0007	−1.2957	0.1951	

Direct Effect of IV on DV controlling for Mediator (c' path)						
DV		Coef. B	Std.error	t-value	p-value	VIF
ROE	*Adjusted R^2: 0.1599* Constant	35.6065	13.8458	2.5716	0.0101	
	TADR	8.6385	15.4499	0.5591	0.5761	2.609
	TDTC BV	−8.4572	0.2810	−30.0932	0.0000	8.283
	TDTC MV	67.0581	17.1996	3.8988	0.0001	6.405
	LTDTC BV	724.8983	19.6970	36.8025	0.0000	3.454
	LTDTC MV	−39.5614	8.3533	−4.7360	0.0000	2.435
	STDTC BV	390.9616	17.9440	21.7879	0.0000	2.952
	STDTC MV	15.6664	14.7049	1.0654	0.2867	5.508
	Firm size (log total assets)	−32.6298	6.4101	−5.0903	0.0000	19.920
	Firm size (log total sales)	27.6560	6.0132	4.5992	0.0000	19.436
	Economic growth (GDP)	4.5952	0.6626	6.9356	0.0000	1.327
	Economic growth (GDI)	−0.5682	0.4084	−1.3913	0.1642	1.353

Part 2: No multi-collinearity (dropping log total sales)						
IV to Mediators (*a* paths)						
Mediator	X to M	Coef. B	Std.error	t-value	p-value	Adj. R^2
TDTC MV	Constant	0.7477	0.0203	36.8486	0.0000	0.0090
	Firm size (log total assets)	0.0196	0.0024	**8.0772**	**0.0000**	
	Economic growth (GDP)	−0.0022	0.0011	−2.0521	0.0402	
	Economic growth (GDI)	−0.0001	0.0007	−0.1673	0.8671	
STDTC MV	Constant	0.7394	0.0221	33.5251	0.0000	0.0088
	Firm size (log total assets)	0.0202	0.0026	7.6672	0.0000	
	Economic growth (GDP)	−0.0024	0.0012	−2.0592	0.0395	
	Economic growth (GDI)	−0.0014	0.0007	**−1.9585**	0.0502	

(continued)

Table 6.2 (continued)

Part 3: No multi-collinearity (dropping log total assets)						
IV to Mediators (*a* paths)						
Mediator	X to M	Coef. B	Std.error	t-value	p-value	Adj. R^2
TDTC MV	Constant	0.7579	0.0197	38.5097	0.0000	0.0086
	Firm size (log total sales)	0.0181	0.0023	**7.8948**	**0.0000**	
	Economic growth (GDP)	−0.0024	0.0011	−2.1873	0.0288	
	Economic growth (GDI)	0.0000	0.0007	0.0684	0.9455	
STDTC MV	Constant	0.7338	0.0213	34.4006	0.0000	0.0099
	Firm size (log total sales)	0.0206	0.0025	8.3052	0.0000	
	Economic growth (GDP)	−0.0025	0.0012	−2.0914	0.0365	
	Economic growth (GDI)	−0.0011	0.0007	−1.5394	0.1237	

significant (t = −1.2957) but it is significant in Part 2 (t = −1.9585). In general, the presence of multi-collinearity not only complicates the interpretation of results, but also makes the confirmation of the effect of any single variable more problematic because of their interrelationships Hair et al. (2010).

In addition, the results for significant mediation effects reveal a contradiction between the regression approach (Sobel test) based on Baron and Kenny (1986) and the bootstrap value of 95% BCa CI. For example, the result of part 2, the Sobel test analysis for criterion (DV) = ROE, predictors (X) = firm size (log total assets), GDP, GDI, mediator (M) = TDTC-BV shows that all predictors of the criterion variables significantly mediate through TDTC-BV, however, the bootstrap values of 95% BCa CI do not show any significant mediation effect. This again as implied in Study 1, the results from the two significance estimates (Sobel test and bootstrap) appear to be slightly different. As in Study 1, it is hard to determine which predictor and criterion variables are mediated by firm leverage. It is probably because the regression approach based on Baron and Kenny (1986) does not involve a replacement procedure (without resampling). So, when bootstrapping is used to test for mediation effects (95% BCa CI), the values of β and the standard errors will differ from the original value (without resampling). Thus, estimating the t-statistics of Sobel and the 95% BCa CI leads to this contradiction. The mediation effects become weaker when complications arise in explaining one to one items of particular attributes of interest. In any scientific undertaking, in statistical analysis the convenient approach is a central consideration for most researchers. Thus, this approach to determine mediation is not as easy as it looks especially when arguments and contradictions still abound.

PLS-SEM

In the PLS-SEM approach all the construct measures are dealt with in one simultaneous model. It is well-known that one advantage of structural equation modelling is being able to work with multiple constructs. Case study 2 takes this opportunity to examine the mediation effects of firm leverage on firm financial performance using

Table 6.3 The measurement model

Attributes	AVE	Composite reliability (CR))	R square (R^2)
Economic growth	0.7061	0.8241	0
Firm size	0.986	0.9929	0
Firm leverage	0.5773	0.9042	0.0701
Firm financial performance	0.5966	0.7875	0.1044

two capital structure determinants. Tables 6.3, 6.5 and 6.6 validate the multi-item scales that have strong measures in the PLS-SEM approach. The first estimates of the elements of the factorial construct validity in the measurement model present a highly valid statistical point. In Table 6.3, most of the convergent validity, i.e., Average Variance Extracted (AVE) values are above 0.5 and likewise, the composite reliability (CR) value for most constructs is at least 0.7[15]. The model statistically supports the reliability and convergent validity for the measurement model between the LVs and their indicators (see examples from Henseler et al. 2017; Hair et al. 2017; Rigdon 2013; Sarstedt et al. 2017). This distinction inspires confidence in using SEM over regression analysis. Next we proceed to test the statistical significance of the parameter estimates. The PLS-SEM analysis offers an overwhelming advantage when working with multi-item scales rather than single-item scales (Study 1) because multi-scales particularly involve convergent validity in the measurement model assessment. A single-item scale is measured with one indicator to one particular latent variable thus the single item could not gauge convergent validity (i.e., Average Variance Extracted (AVE) and construct reliability (CR)). In short, the AVE and CR are derived by estimating the share among other indicators (i.e., two or more indicators) on the particular latent variable (LVs). Thus, single-item scales cannot be appraised because AVE calculation needs the weight of the factor loading for all possible indicators (i.e., two or more) for the particular LVs.

Factor loading (λ) is the combination of a number of individual items or indicators into a single composite measure Hair et al. (2010). The higher the degree of the factor loading the higher is the connection between the variables and factors. The ultimate goal of factor loading is to avoid an individual item not representing the exact facet of the concept. According to Hair et al. (2010), combining several items or indicators that have been measured in the same concept into a single variable (i.e., LVs) enhances the reliability of the multivariate measurements. Thus, it is essential to determine the summated scale of factor loadings before conducting further analysis. The guidelines proposed by Hair et al. (2010) in assessing factor loadings are as follows in Table 6.4:

[15]AVE is the parameter to measure convergent validity that estimates the "degree to which two measures of the same concept are correlated (Hair et al. 2013)". Specifically, it measures the degree of multiple items in the same construct. The cut-off value for good convergent validity in AVE is set as 0.5 or above. That score of 0.5 means 50% of the measurement variance is accounted for (Fornell and Larcker 1981; Hair et al. 2013). Composite reliability (CR) is assessed by means of all the indicators assigned that have strong mutual correlations to the same construct. So, CR is used to check how well all indicators are connected with the construct. The acceptable threshold cut-off value for CR is 0.7 or above. But cut-off values of 0.5 and 0.6 have also been acceptable (Bagozzi and Yi 1988; Hair et al. 2013).

Table 6.4 Guidelines for identifying significance factor loading

Range of factor loading	Significance
Loading of ± 0.30 to 0.40	It is considered to meet the minimal level for structure interpretation
Loading of ± 0.50 or greater	It is considered practically significant
Loading of ± 0.70 or greater	It is considered indicative of a well-defined structure of any factor analysis

Table 6.5 Statistically significant value (the structural model)

Path coefficient	Coeff. β	Std. error	95% BCa CI
Economic growth → firm leverage	−0.0696	0.0129	(0.000, −0.055)
Economic growth → firm financial performance	0.0982	0.0142	(0.002, 0.124
Firm size → firm leverage	0.2573	0.0122	(0.000, 0.306)
Firm size →firm financial performance	0.1616	0.0255	(0.005, 0.181)
Firm leverage → firm financial performance	−0.2979	0.0282	(0.006, −0.239)

Table 6.6 Mediation test analysis results

Path coefficient	Indirect effect	95% BCa CI	Results
Firm size → firm leverage → firm performance	0.000	(0.002, 0.067)	Accepted
Economic growth → firm leverage → firm performance	0.000	(0.000, 0.028)	Accepted

The mediating relationship effects of firm leverage (M) between capital structure determinants and the firm financial performance in PLS–SEM

Measurement results of the factorial construct validity for the endogenous and exogenous variables hypothesised to influence the determinants of capital structure and firm' financial performance estimated using Partial Least Square in variance based Structural Equation Modelling (PLS-SEM). The measurement estimates are calculated by the PLS algorithm with the path weighting scheme, Mean 0, Variance 1, Maximum iteration 300.

Figure 6.11 shows that the structure among the latent variables and its formed indicators are highly loaded (λ). The summated scale for each indicator or item on its attributes (LVs) mostly average higher than λ = 0.7, which is considered high as a measure of reliability. In other words, the variable structures are well-defined and better-rounded. It can be said that this is one of the strength of multi-item scales in multivariate analysis because it provides greater confidence and accuracy in the measurement of the selected indicator's attributes (LVs). In addition, the R-square (R^2) values show that the total proportion of the outcome variation is slightly larger than in the single items because of the weighted variables in the simultaneous method.

The table demonstrates the estimated statistical significance of Partial Least Square in variance based Structural Equation Modelling (PLS-SEM) for the determinants of capital structure and firm' financial performance. This table also expresses the structural path correlation coefficient (β) and standard error among

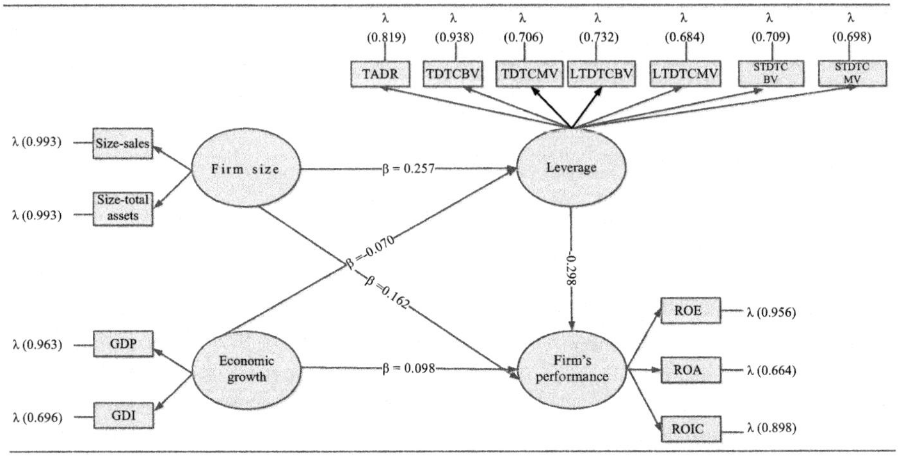

Fig. 6.11 A measurement model using PLS-SEM

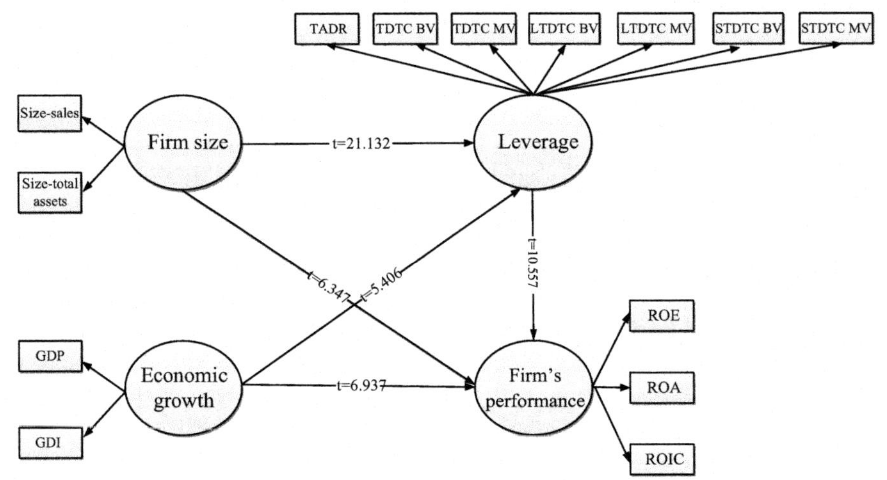

Fig. 6.12 A structural model using PLS-SEM

hypothesized exogenous variables with the firm leverage and the firm financial performance with the capital structure attributes. The PLS path model measures the statistically significant value and the Beta (β) coefficient by using resampling from the bootstrapping procedures for a number of samples of 5000 for all sample; total population estimates with the number of cases (N) = 7819

The 5000 bootstrap sample with 7819 cases is considered desirable for statistical structural path models. All path coefficients in Fig. 6.12 are statistically significant

(Table 6.5 for a summary of the t-statistics). For instance, leverage is significantly related to firm financial performance is significant (i.e., 95% BCa CI of 0.006 (bootLLCI) to −0.239 (bootULCI)).When multi-item scales were used (for all constructs), the predicted measures are very strong.

Table 6.6 shows that firm leverage significantly mediates the relationship between X (firm size and economic growth) and Y (firm financial performance). Similar to findings from the PLS-SEM analysis in Study 1, the 95% BCa bootstrap CI provides a conclusion consistent with the bootstrap statistics estimation. This indicates superiority of the PLS-SEM over regression analysis when we used the 95% BCa bootstrap CI technique. Study 2 also shows consistent results when the models separately analyse two capital structure factors, i.e., firm size and economic growth. That is not our focus so the results are not provided here.

However, multi scale models also have its limitations. Simultaneous models with multi-item scales could result in researchers having difficulty in identifying which individual indicators are significant. For example, the results of the analysis might find that firm leverage mediates relationships between capital structure determinants and firm financial performance. However, if one has seven firm leverage measures it would be difficult to identify which ones are significant. For example we would not know if TADR, TDTC book or market firm leverage or other measures are significant. This sort of question is better answered by the single simultaneous model in Study 1, which also uses PLS-SEM. In a single simultaneous model (one item to one construct) one could investigate in more detail the effect of each individual variable. However, if the focus of the study is on causal relationships, the simultaneous estimation is better.

In conclusion, Study 2 illustrates that when aggregated all multi-scales are aggregated in one model via PLS-SEM, the significance of mediation effects is strongly detectable. We find that the results of both studies 1 and 2 indicate that PLS-SEM is superior to regression analysis especially by using the 95% BCa bootstrap CI estimation. The structural equation in PLS-SEM has shown a consistent advantage for distinguishing mediation relationships.

6.5 Discussion

The objectives of this study are (i) to identify the limitation, weaknesses and strengths of the regression and PLS-SEM approaches and, (ii) compare the predictions of the two approaches. In furtherance of the above objectives, two case studies on capital structure are conducted. The case studies compare the ability of PLS-SEM and regression analysis in detecting mediation effects. In the first case study, each construct (LV) is analysed separately using regression and PLS-SEM techniques. The second case study conducts the analysis using multiple indicators for each construct. The objectives can be discussed as follows: When hypothesising mediation or causal relationships, any simultaneous equation model (SEM) is the most general tool. It has been shown that PLS-SEM is better than traditional regression analysisespecially by using the 95% BCa bootstrap CI estimation.

Mediation models are either simple or multiple-item scale models. Their consistency in terms of statistical significance as well as the factorial construct validity in PLS-SEMs is an advantage over the regression technique. PLS-SEM can easily handle problems of multi-collinearity relative toregression analysis (specifically, when dealing with multiple variables with the possibility that any of the independent variables could be correlated among themselves). A specific mediation/indirect effect represents the ability of firm leverage "M" to mediate such a relationship after controlling all other independent variables. These effects of mediation relationships will normally, but not inevitably, be weakened by co-llinearity or redundancy among the predictor variables. This phenomenon is not necessarily a problem, but it could lead the researcher to conclude erroneously that firm leverage is a mediator when in fact it is not, or that firm leverage is not a mediator when it is. PLS-SEM has proven the strength to minimize this problem; this is not so for regression.

Another point concerns the assumption of normality of the sampling distributions. Traditional regression analysis i.e., ordinary least squares (OLS), is parametric but PLS-SEM regression is non-parametric. Regression analysis requires variables to be normally distributed, while PLS-SEM does not require any distributional form for the measured variables (Noonan and Wold 1986; Lohmöller 1989; Wold 1982). That is, PLS-SEM relies on a non-parametric bootstrap procedure to test the coefficients for its significance level. In regression, if all assumptions are met, then the parameter estimations tend to be optimal[16] (specifically, the best linear unbiased estimates "BLUE"). If the datasets all meet the assumptions, the regression estimates are optimal. If the assumptions are nearly met, the regression estimates are nearly optimal. However, PLS-SEM has no presumption of any distributional form thus mathematically the estimates are not expected to be optimal and cannot be optimal regardless of what assumption are made (Goodhue et al. 2012; Rönkkö et al. 2015). However, most PLS-SEM researchers expect that the estimates from PLS are mathematically optimal regardless of the sample distribution[17]. Goodhue et al. (2012) stated that PLS-SEM has less statistical efficacy if we do not assume a normal distribution for our variables. Therefore, this leads to a weakness or limitation of PLS-SEM when the data do not conform to a certain distributional requirement compared with other techniques such as regression. Thus, when a non-parametric method makes fewer assumptions regarding the nature of a

[16]The regression assumptions are: (i) the relationship of Xs and Y is linear, (ii) "non-stochastics" of Xs and no exact linear relationship exists between any two Xs (i.e., Xs is known with certainty which is the Xs are measured without errors), (iii) the value of error term are expected value of zero, having a constant variance for all observations, uncorrelated across observations and normally distributed (see the Gauss-Markov theorem).

[17]The case of assume the optimal exists is when developer of PLS-SEM started to asserted such as, Wold (1982, p. 28) says "scarifying optimally", "the gains of sacrificing optimally are considerable".. "overall optimization property"… "reliability".. etc.

distribution, the approach is usually less powerful than its parametric counterparts. However, if the assumptions of the parametric test are not met, then parametric tests can have relatively less power. In the case of the parametric method, the assumptions are not always met thus a shift to a nonparametric test sometimes would be more proper and can be more powerful (as the case of the PLS-SEM technique). However, the assumption of normality is required for many statistics in the behavioural sciences so PLS-SEM has a strong advantage regarding this issue. Iacobucci et al. (2007) claimed that PLS-SEM is also expected to provide a positive finding in simple mediation models even for small samples ($n < 200$). Hence, this technique is suitable for any data analysis. However, PLS-SEM gives more reliable and accurate result with a large sample (Jannoo et al. 2014; Latan and Noonan 2017; Lee et al. 2011; Marcoulides and Saunders 2006), which is the case in this study.

The case study presented here not only involved measured variables (single-item scales) with no correction for unreliability but also involved multi-item scales, and even latent variables in PLS-SEM. This extension of latent variables is one of the most fascinating features of PLS-SEM that allow the inclusion of measurement error, unlike regression analysis which does not. Rather than hypothesising imperfect individual indicators, PLS-SEM is able to test a hypothesis using the latent variable -construct that can be measured by a number of observable indicators. Note too, that PLS-SEM can accommodate many constructs easily, such as:$X_1 \rightarrow Y$, $X_1 \rightarrow M$, $M \rightarrow Y$, $X_2 \rightarrow Y$, $X_2 \rightarrow M$, $M \rightarrow Y$, *etc.*Therefore, PLS-SEM is recommended for exploring causal relationships especially in mediating with reasonable control of measurement errors. This is also the power of PLS-SEM. However, there is a limitation in multi-item scales with regards to the latent variables in PLS-SEM. This limitation would probably embrace the practical implication for each item/indicator. This means that, for example, sometimes in a capital structure study a researcher would like to examine how far each individual firm leverage measure acts as mediator (i.e., is it TADR, TDTC book or market leverage or other measures?). The researcher can still solve this matter by using the PLS-SEM approach, but by using a single simultaneous model (single-item scales), as in case Study 1.

This discussion about PLS-SEM and macro-regressions presented here applies only to the case of simple and multiple-item scale models. Many extensions from simple to complex models such as multiple mediation or moderated mediation are possible (see in Fig. 6.8). This, in turn, can present more complex models, such as those with multiple mediators that link concepts differently. We also discussed the Sobel test and the bootstrapping technique (i.e., 95% BCa bootstrap CI) both used to evaluate mediation effects in the current regression version of macro syntax as well as the 95% BCa bootstrap CIin PLS-SEMs. The regression macro syntax described in this chapter provides users with tools to analyse interesting mediation effects using SPSS with a command "indirect" and "mediate". However, researchers should be aware of the contradictory results that appear from the Sobel test and the bootstrap techniquewith this macro as noted in the previous section. Table 6.7 shows the summary of the capabilities of the statistical analysis techniques in the context of capital structured study.

Table 6.7 Capabilities of the statistical analysis techniques in the context of capital structure

Capabilities	PLS-SEM	Regression
Examines multiple indicators for each construct	Supported	Not supported
The consistency in terms of statistical significance as well as the factorial construct validity	Supported	Not readily supported
Can easily handle problems of multi-collinearity	Supported	Not readily supported
Minimize the problem of co-linearity or redundancy among the predictor variables	Supported	Not supported
Normal distribution	Not supported	Supported
Can cope with relatively small sample size	Supported	Supported
Test hypothesis using latent variables	Supported	Not supported
Can explain the practical implication for each of the multi-scale of latent variable	Not readily supported	Not supported
Analysis of single items loading paths	Supported	Not supported
Maps paths to many dependent (latent or observed) variables in the same research model and analyse all the paths simultaneously rather than one at a time	Supported	Not supported
Maps specific and error variance of the observed variables into the research model	Not supported	Not supported
Analyses all the paths, both measurement and structural, in one analysis.	Supported	Not supported

6.6 Summary

The empirical evidence and statistical theory from the study can be summarized as follows. First, before conducting mediation analysis, the source of mediation should be clearly understood in terms of how it can influence the direct cause and effect between X (predictor variable) and Y (criterion). Second, the model fit via PLS-SEM appears to be superior to ordinary regression analysis. The measurement model of simultaneous structural analysis in PLS-SEM for direct and indirect effects should be fitted into one model (see Fig. 6.11 of Study 2). Third, the significance of the mediation effect will be evident when both path coefficients $X \rightarrow M$ and $M \rightarrow Y$ are significant. If both paths are not significant, an analyst can stop the analysis with the conclusion of no mediation effect. To determine the significance of the direct path of $X \rightarrow Y$a comparative t-test such as the Sobel test or bootstrapping should be

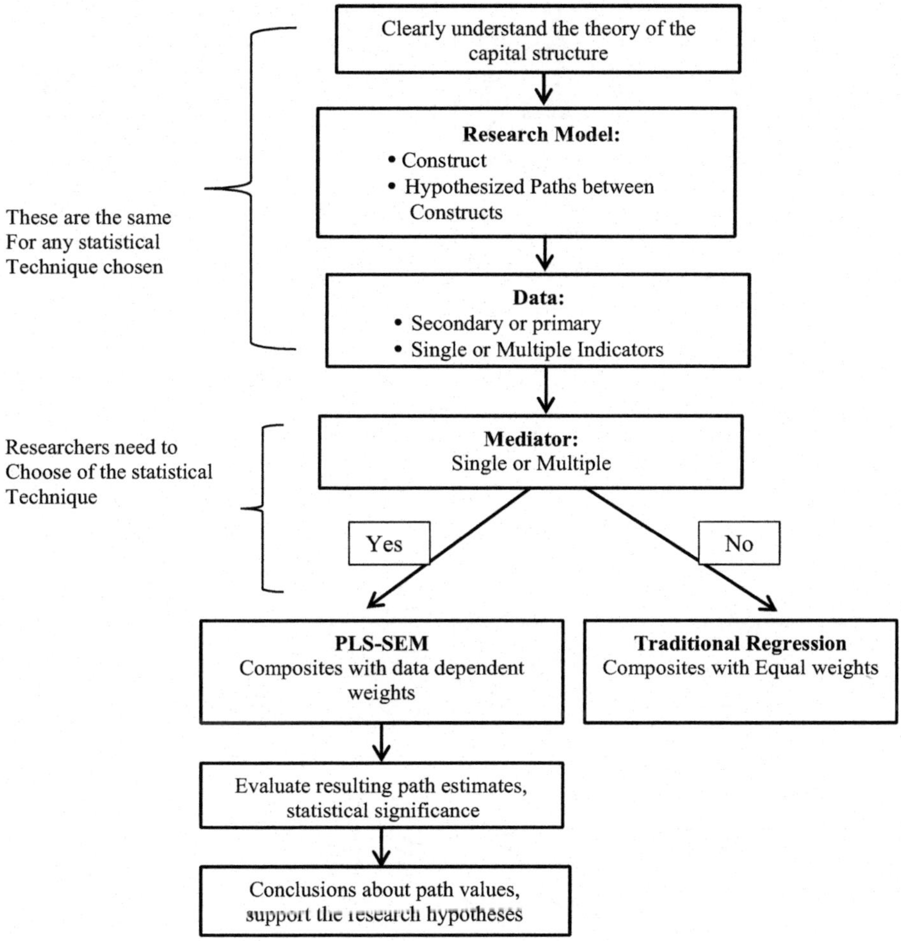

Fig. 6.13 The choice of statistical techniques in the context of capital structure

applied. This is because, even though the path coefficient $X \rightarrow Y$ is not different from zero (insignificant), it might strengthen the indirect path causing it to be significant. The summary of the choice of statistical techniques in the context of capital structure perspective can be seen in Fig. 6.13.

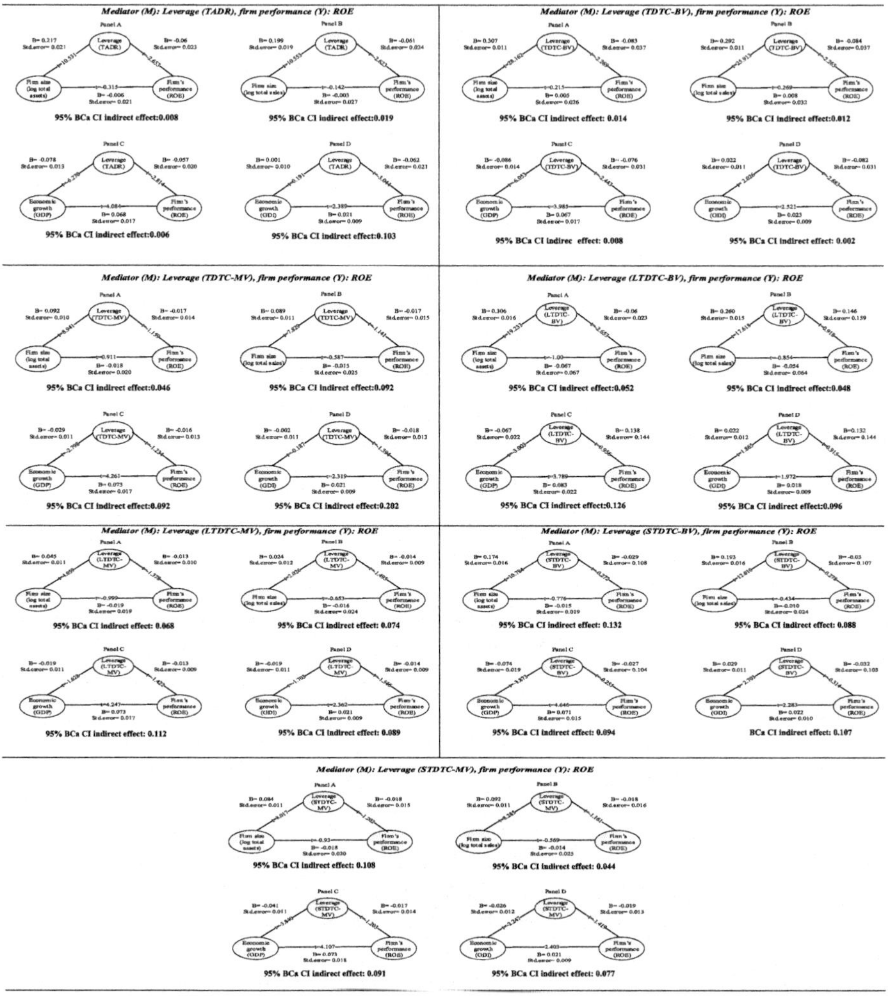

Fig. 6.14 Single equation model

Appendix 6.1

6.4.1. Study 1

Table 6.8 Maco syntax matrix regression procedures

IV to Mediators (a paths)							Direct effect of IV on DV controlling for mediator (c' path)						
Model	Mediator	X to M	Coef. B	Std. error	t-value	p-value	Model	Mediator		Coef. B	Std. error	t-value	p-value
A	TADR	Firm size (log total assets)	0.0324	0.0016	19.6983	0.000	A	TADR	Firm size (log total assets)	−0.9194	1.5789	−0.5823	0.5604
		Firm size (log total sales)	0.0283	0.0016	18.021	0.000			Firm size (log total sales)	−0.4956	1.4988	−0.3307	0.7409
		Economic growth (GDP)	−0.0047	0.0007	−6.9892	0.000			Economic growth (GDP)	3.8027	0.624	6.0938	0.000
		Economic growth (GDI)	0.0001	0.0004	0.1709	0.8643			Economic growth (GDI)	0.7146	0.3794	1.8833	0.0597
B	TDTC BV	Firm size (log total assets)	4.4833	0.157	28.5505	0.000	B	TDTC BV	Firm size (log total assets)	0.7755	1.6172	0.4795	0.6316
		Firm size (log total sales)	4.0557	0.1503	26.9778	0.000			Firm size (log total sales)	1.1021	1.5333	0.7188	0.4723
		Economic growth (GDP)	−0.5051	0.0665	−7.5934	0.000			Economic growth (GDP)	3.6913	0.6236	5.9195	0.000
		Economic growth (GDI)	0.0807	0.0406	1.987	0.047			Economic growth (GDI)	0.7727	0.379	2.039	0.0415
C	TDTC MV	Firm size (log total assets)	0.0194	0.0024	8.1797	0.000	C	TDTC MV	Firm size (log total assets)	−2.5467	1.5549	−1.6379	0.1015
		Firm size (log total sales)	0.0179	0.0023	7.9336	0.000			Firm size (log total sales)	−1.9195	1.4808	−1.2963	0.1949
		Economic growth (GDP)	−0.0025	0.001	−2.6023	0.0093			Economic growth (GDP)	4.0248	0.6238	6.4526	0.000
		Economic growth (GDI)	−0.0001	0.0006	−0.1835	0.8544			Economic growth (GDI)	0.709	0.3804	1.8635	0.0624

(continued)

Table 6.8 (continued)

IV to Mediators (a paths)							Direct effect of IV on DV controlling for mediator (c' path)						
Model	Mediator	X to M	Coef. B	Std. error	t-value	p-value	Model	Mediator		Coef. B	Std. error	t-value	p-value
D	LTDTC BV	Firm size (log total assets)	0.0412	0.0014	28.4333	0.000	D	LTDTC BV	Firm size (log total assets)	−9.1094	1.6046	−5.6771	0.000
		Firm size (log total sales)	0.0333	0.0014	23.8295	0.000			Firm size (log total sales)	−7.0529	1.5086	−4.675	0.000
		Economic growth (GDP)	−0.0036	0.0006	−5.9077	0.000			Economic growth (GDP)	4.5583	0.6186	7.3684	0.000
		Economic growth (GDI)	0.0007	0.0004	1.9361	0.0529			Economic growth (GDI)	0.6134	0.3771	1.6266	0.1039
E	LTDTC MV	Firm size (log total assets)	0.0121	0.003	4.0071	0.000	E	LTDTC MV	Firm size (log total assets)	−2.6674	1.5484	−1.7227	0.085
		Firm size (log total sales)	0.0061	0.0029	2.1261	0.0335			Firm size (log total sales)	−2.0692	1.474	−1.4039	0.1604
		Economic growth (GDP)	−0.002	0.0012	−1.6336	0.1024			Economic growth (GDP)	4.0383	0.6234	6.4773	0.000
		Economic growth (GDI)	−0.0013	0.0007	−1.6954	0.09			Economic growth (GDI)	0.7012	0.3804	1.8432	0.0653
F	STDTC BV	Firm size (log total assets)	0.024	0.0015	15.689	0.000	F	STDTC BV	Firm size (log total assets)	−2.1015	1.5845	−1.3263	0.1848
		Firm size (log total sales)	0.0253	0.0015	17.4705	0.000			Firm size (log total sales)	−1.3949	1.5124	−0.9223	0.3564
		Economic growth (GDP)	−0.0041	0.0006	−6.5585	0.000			Economic growth (GDP)	3.9439	0.6262	6.2984	0.000
		Economic growth (GDI)	0.001	0.0004	2.6383	0.0083			Economic growth (GDI)	0.7451	0.3815	1.9532	0.0508

IV to Mediators (a paths)							Direct effect of IV on DV controlling for mediator (c' path)						
Model	Mediator	X to M	Coef. B	Std. error	t-value	p-value	Model	Mediator		Coef. B	Std. error	t-value	p-value
G	STDTC MV	Firm size (log total assets)	0.0195	0.0026	7.5118	0.000	G	STDTC MV	Firm size (log total assets)	−2.6123	1.5693	−1.6646	0.096
		Firm size (log total sales)	0.0203	0.0025	8.2431	0.000			Firm size (log total sales)	−1.9423	1.4929	−1.3011	0.1933
		Economic growth (GDP)	−0.0037	0.001	−3.5868	0.0003			Economic growth (GDP)	4.0153	0.6254	6.4205	0.000
		Economic growth (GDI)	−0.0015	0.0006	−2.3065	0.0211			Economic growth (GDI)	0.6952	0.3818	1.8208	0.0687

Direct effects of mediators on DV (b paths)					Total effect of IV on DV (c path)				
M to Y	Coef. B	Std.error	t-value	p-value	X to Y	Coef. B	Std. error	t-value	p-value
TADR	−56.6422	10.5804	−5.3535	0.0000	Firm size (log total assets)	−2.73	1.544	−1.769	0.0770
TDTC BV	−0.782	0.1108	−7.0552	0.0000	Firm size (log total sales)	−2.099	1.471	−1.427	0.1540
TDTC MV	−10.7824	7.3952	−1.458	0.1449	Economic growth (GDP)	4.051	0.623	6.502	0.0000
LTDTC BV	154.9192	11.9324	12.9831	0.0000	Economic growth (GDI)	0.711	0.38	1.869	0.0620
LTDTC MV	−6.7315	5.8122	−1.1582	0.2468					
STDTC BV	−29.1996	11.5452	−2.5291	0.0115					
STDTC MV	−10.7524	6.8313	−1.574	0.1155					

	Mediation effects		Percentage of the total effect	Bias corrected confidence intervals		Bootstrap results for mediation effects
	Sobel test	Goodman test		Lower	Upper	
Sizeta → TADR → ROE	−5.0957	−5.0895	66.3305	−5.269	−0.5492	Support for mediation
Sizesales → TADR → ROE	−5.1318	−5.1247	76.3888	−4.6141	−0.5123	Support for mediation
GDP → TADR → ROE	4.0932	4.066	6.1302	0.0684	0.6876	Support for mediation
GDI → TADR → ROE	−0.1708	−0.1681	−0.5716	−0.073	328	No support for mediation

(continued)

Table 6.8 (continued)

	Mediation effects		Percentage of the total effect	Bias corrected confidence intervals		Bootstrap results for mediation effects
	Sobel test	Goodman test		Lower	Upper	
Sizeta → TDTC BV → ROE	−6.8492	−6.8453	128.3982	−8.2226	−1.3988	Support for mediation
Sizesales → TDTC BV → ROE	−6.9112	−6.9067	152.5059	−7.3196	−1.1653	Support for mediation
GDP → TDTC BV → ROE	5.0416	5.0173	8.8805	0.1124	1.0321	Support for mediation
GDI → TDTC BV → ROE	−1.9174	−1.9009	−8.7543	−0.2512	−0.013	Support for mediation
Sizeta → TDTC MV → ROE	−1.2634	−1.2543	7.6537	−0.3303	−0.1514	Support for mediation
Sizesales → TDTC MV → ROE	−1.3315	−1.3214	9.4334	−0.297	−0.1337	Support for mediation
GDP → TDTC MV → ROE	1.2529	1.1873	0.6479	0.0103	0.053	Support for mediation
GDI → TDTC MV → ROE	0.224	0.1907	0.1793	−0.0121	0.0162	No support for mediation
Sizeta → LTDTC BV ->ROE	11.811	11.8049	−233.5834	−2.4789	24.4326	No support for mediation
Sizesales → LTDTC BV → ROE	11.1461	11.1384	−235.9986	−1.989	20.1296	No support for mediation
GDP → LTDTC BV → ROE	−5.3272	−5.3129	−12.5172	−1.8063	0.2173	No support for mediation
GDI → LTDTC BV → ROE	1.9135	1.9068	13.6538	−0.0315	0.7993	No support for mediation
Sizeta → LTDTC MV → ROE	−0.8669	−0.843	2.9747	−0.2293	−0.0116	Support for mediation
Sizesales → LTDTC MV → ROE	−0.7303	−0.6759	2.0349	−0.1603	−0.0054	Support for mediation
GDP → LTDTC MV → ROE	0.9142	0.8161	0.3187	0.0000	0.0451	Support for mediation
GDI → LTDTC MV → ROE	1.0405	0.9376	1.2344	0.0000	0.0418	Support for mediation
Sizeta → STDTC BV → ROE	−2.2382	−2.2338	25.7008	−4.4546	2.4637	No support for mediation
Sizesales → STDTC BV → ROE	−2.3695	−2.3658	35.9862	−5.0232	2.4132	No support for mediation
GDP → STDTC BV → ROE	2.1745	2.1524	2.6783	−0.3848	0.8552	No support for mediation
GDI → STDTC BV → ROE	−2.0551	−1.9905	−4.5947	−0.2356	0.0958	No support for mediation
Sizeta → STDTC MV ->ROE	−0.8695	−0.8622	7.6794	−0.3662	−0.1322	Support for mediation
Sizesales → STDTC MV ->ROE	−1.1065	−1.0988	10.5401	−0.3558	−0.1413	Support for mediation
GDP → STDTC MV → ROE	1.3019	1.2606	0.9183	0.0165	0.0666	Support for mediation
GDI → STDTC MV → ROE	1.2411	1.1711	2.3517	0.0033	0.0424	Support for mediation

References

Baron, R. M., & Kenny, D. A. (1986). The moderator–mediator variable distinction in social psychological research: Conceptual, strategic, and statistical considerations. *Journal of Personality and Social Psychology, 51*(6), 1173.

Bennett, M., & Donnelly, R. (1993). The determinants of capital structure: Some UK evidence. *The British Accounting Review, 25*(1), 43–59.

Bevan, A. A., & Danbolt, J. (2002). Capital structure and its determinants in the UK—a decompositional analysis. *Applied Financial Economics, 12*(3), 159–170.

Cepeda, G., Nitzl, C., & Roldán, J. L. (2017). Mediation analyses in partial least squares structural equation modeling: Guidelines and empirical examples. In H. Latan & R. Noonan (Eds.), *Partial least squares path modeling: Basic concepts, mhodological issues and applications* (pp. 173–195). Cham: Springer International.

Chang, C., Lee, A. C., & Lee, C. F. (2009). Determinants of capital structure choice: A structural equation modeling approach. *The Quarterly Review of Economics and Finance, 49*(2), 197–213.

Chiarella, C., Pham, T. M., Sim, A. B., & Tan, M. M. L. (1991). Determinants of corporate capital structure: Australian evidence. In G. S. Rhee & R. P. Chan (Eds.), *Pasific basin capital markets research* (Vol. 3, pp. 139–158). Amsterdam: North Holland.

Chin, W. W. (2010). How to write up and report PLS-SEM analyses. In V. E. Vinzi, W. W. Chin, J. Henseler, & H. Wang (Eds.), *Handbook of partial least squares: Concepts, methods and applications in marketing and related fields*. Berlin: Springer.

Goodhue, D. L., Lewis, W., & Thompson, R. (2012). Does PLS have advantages for small sample size or non-normal data? *MIS Quarterly, 36*(3), 981–1001.

Hair, J. F., Black, W. C., Babin, B. J., & Anderson, R. E. (2010). *Multivariate data analysis* (7th ed.). Upper Saddle River: Prentice-Hall.

Hair, J. F., Sarstedt, M., Ringle, C. M., & Mena, J. A. (2012). An assessment of the use of partial least squares structural equation modeling in marketing research. *Journal of Academy of Marketing Research, 40*(3), 414–433.

Hair, J. F., Hult, G. T. M., Ringle, C., & Sarstedt, M. (2013). *A primer on partial least squares structural equation modeling (PLS-SEM)*. Thousand Oaks: Sage.

Hair, J. F., Hollingsworth, C. L., Randolph, A. B., & Chong, A. Y. L. (2017). An updated and expanded assessment of PLS-SEM in information systems research. *Industrial Management & Data Systems, 117*(3), 442–458.

Hair, J. F., Hult, G. T. M., Ringle, C. M., & Sarstedt, M. (2017). *A primer on partial least squares structural equation modeling (PLS-SEM)* (2nd ed.). Thousand Oaks: Sage.

Hayes, A. F. (2009). Beyond Baron and Kenny: Statistical mediation analysis in the new millennium. *Communication Monographs, 76*(4), 408–420.

Hayes, A. F. (2013). *Introduction to mediation, moderation, and conditional process analysis: A regression-based approach*. New York: Guilford Press.

Henseler, J. (2010). On the convergence of the partial least squares path modeling algorithm. *Computational Statistics, 25*(1), 107–120.

Henseler, J., Hubona, G., & Ray, P. A. (2017). Partial least squares path modeling: Updated guidelines. In H. Latan & R. Noonan (Eds.), *Partial least squares path modeling: Basic concepts, methodological issues and applications* (pp. 19–39). Cham: Springer International.

Iacobucci, D., & Duhachek, A. (2003). Advancing alpha: Measuring reliability with confidence. *Journal of Consumer Psychology, 13*(4), 478–487.

Iacobucci, D., Saldanha, N., & Deng, X. (2007). A meditation on mediation: Evidence that structural equations models perform better than regressions. *Journal of Consumer Psychology, 17*(2), 139–153.

Jairo, I. (2009). The use of structural equation modelling (SEM) in capital structure empirical analysis. *KCA Journal of Business Management, 1*(1), 11–35.

Jannoo, Z., Yap, B., Auchoybur, N., & Lazim, M. (2014). The effect of nonnormality on CB-SEM and PLS-SEM path estimates. *International Journal of Mathematical, Computational, Physical and Quantum Engineering, 8*(2), 285–291.

Jose, P. E. (2013). *Doing statistical mediation and moderation*. New York: Guilford Press.

Judd, C. M., & Kenny, D. A. (1981). Process analysis estimating mediation in treatment evaluations. *Evaluation Review, 5*(5), 602–619.

Latan, H. (2015). *Applied statistical data analysis with IBM SPSS*. Bandung: Alfabeta.

Latan, H., & Noonan, R. (Eds.). (2017). *Partial least squares path modeling: Basic concepts, methodological issues and applications*. Cham: Springer International.

Lee, L., Petter, S., Fayard, D., & Robinson, S. (2011). On the use of partial least squares path modeling in accounting research. *International Journal of Accounting Information Systems, 12*(4), 305–328.

Lin, F.-J. (2008). Solving multicollinearity in the process of fitting regression model using the nested estimate procedure. *Quality & Quantity, 42*(3), 417–426.

Lohmöller, J.-B. (1989). *Latent variable path modeling with partial least squares*. Berlin: Springer.

MacKinnon, D. P. (2000). Contrasts in multiple mediator models. In J. S. Rose & L. Chassin (Eds.), *Multivariate applications in substance use research: New methods for new questions*. Mahwah: Erlbaum.

MacKinnon, D. P., Warsi, G., & Dwyer, J. H. (1995). A simulation study of mediated effect measures. *Multivariate Behavioral Research, 30*(1), 41–62.

MacKinnon, D. P., Lockwood, C. M., Hoffman, J. M., West, S. G., & Sheets, V. (2002). A comparison of methods to test mediation and other intervening variable effects. *Psychological Methods, 7*(1), 83.

MacKinnon, D. P., Lockwood, C. M., & Williams, J. (2004). Confidence limits for the indirect effect: Distribution of the product and resampling methods. *Multivariate Behavioral Research, 39*(1), 99–128.

Marcoulides, G. A., & Saunders, C. (2006). Editor's comments: PLS: a silver bullet? *Mis Quarterly, 30*(2), iii–iix.

Marsh, P. (1982). The choice between equity and debt: An empirical study. *The Journal of Finance, 37*(1), 121–144.

Neter, John, Wasserman, W., & Kutner, M. H. (1990). *Applied Linear Regression Models*. Homewood, Illinois: Richard D. Irwin, Inc.

Nitzl, C., Roldán, J. L., & Cepeda Carrión, G. (2016). Mediation analysis in partial least squares path modeling: Helping researchers discuss more sophisticated models. *Industrial Management & Data Systems, 119*(9), 1849–1864.

Noonan, R., & Wold, H. (1986). Partial least squares path analysis. In T. Husen & T. N. Postlethwaite (Eds.), *The international encyclopedia of education* (Vol. 7, pp. 3769–3775). Oxford: Pergamon Press.

Ozkan, A. (2001). Determinants of capital structure and adjustment to long run target: Evidence from UK company panel data. *Journal of Business Finance & Accounting, 28*(1-2), 175–198.

Preacher, K. J. (2013). *Calculation for the Sobel test: An interactive calculation tool for mediation tests*. Retrieved January 2, 2013, from http://quantpsy.org/sobel/sobel.htm.

Preacher, K. J., & Hayes, A. F. (2004). SPSS and SAS procedures for estimating indirect effects in simple mediation models. *Behavior Research Methods, Instruments, & Computers, 36*(4), 717–731.

Preacher, K. J., & Hayes, A. F. (2008). Asymptotic and resampling strategies for assessing and comparing indirect effects in multiple mediator models. *Behavior Research Methods, 40*(3), 879–891.

Rigdon, E. E. (2013). Partial least squares path modeling. In G. R. Hancock & R. O. Mueller (Eds.), *Structural equation modeling: A second course* (2nd ed., pp. 81–116). Greenwich: Information Age.

Rönkkö, M., McIntosh, C. N., & Antonakis, J. (2015). On the adoption of partial least squares in psychological research: Caveat emptor. *Personality and Individual Differences, 87*(1), 76–84.

Sarstedt, M., Ringle, C. M., & Hair, J. F. (2017). Partial least squares structural equation modeling. In C. Homburg, M. Klarmann, & A. Vomberg (Eds.), *Handbook of market research* (pp. 1–40). Heidelberg: Springer International.

Shrout, P. E., & Bolger, N. (2002). Mediation in experimental and nonexperimental studies: New procedures and recommendations. *Psychological Methods, 7*(4), 422.

Sobel, M. E. (1982). Asymptotic confidence intervals for indirect effects in structural equation models. *Sociological Methodology, 13*, 290–312.

Sobel, M. E. (1986). Some new results on indirect effects and their standard errors in covariance structure models. *Sociological Methodology, 16*, 159–186.

Titman, S., & Wessels, R. (1988). The determinants of capital structure choice. *The Journal of Finance, 43*(1), 1–19.

Vanderweele, T. J. (2015). *Explanation in causal inference: Methods for mediation and interaction*. Oxford: Oxford University Press.

Wold, H. (1982). Soft modelling: The basic design and some extensions. In K. G. Joreskog & H. O. A. Wold (Eds.), *System under indirect observation (Vol. 2)*. Amsterdam: North-Holland.

Yang, C. C., Lee, C. F., Gu, Y. X., & Lee, Y. W. (2010). Co-determination of capital structure and stock returns—A LISREL approach: An empirical test of Taiwan stock markets. *The Quarterly Review of Economics and Finance, 50*(2), 222–233.

Zhao, X., Lynch, J. G., & Chen, Q. (2010). Reconsidering Baron and Kenny: Myths and truths about mediation analysis. *Journal of Consumer Research, 37*(2), 197–206.

Chapter 7
Management Accounting and Partial Least Squares-Structural Equation Modelling (PLS-SEM): Some Illustrative Examples

Christian Nitzl

Abstract In management accounting research, the usefulness of Partial Least Squares-Structural Equation Modelling (PLS-SEM) is commonly underestimated. However, those specific characteristics of PLS-SEM that remain unrecognized in management accounting research appear to be tailor-made for answering a variety of relevant research questions. PLS-SEM is particularly appropriate for management accounting because such research is often conducted at an exploratory stage; the theoretical basis is often weak; archival data or formative measurements are highly relevant; and the predictive orientation of PLS-SEM offers the potential to answer practical questions. Based on three highly ranked published articles in management accounting, I demonstrate how PLS-SEM can be useful in this field. First, Ittner et al. (Account Rev 72:231 -255, 1997) illustrate how accounting research can use archival data in PLS-SEM; second, Hartmann and Maas (Account Business Res 41:439–458, 2011) construct an example of the exploratory use of PLS-SEM in management accounting; and third, Burkert and Lueg (Manage Account Res 24:3–22, 2013) use hierarchical latent construct in PLS-SEM in their management accounting research.

Keywords Management accounting · Archival data · Exploratory · Hierarchical latent construct

In management accounting research, the usefulness of Partial Least Squares-Structural Equation Modelling (PLS-SEM) is commonly underestimated. However, those specific characteristics of PLS-SEM that remain unrecognized in management accounting research appear to be tailor-made for answering a variety of relevant research questions. PLS-SEM is particularly appropriate for management accounting because such research is often conducted at an exploratory stage; the theoretical basis is often weak; archival data or formative measurements are highly

C. Nitzl (✉)
University of the German Armed Forces Munich, Neubiberg, Germany
e-mail: christian.nitzl@unibw.de

N.K. Avkiran, C.M. Ringle (eds.), *Partial Least Squares Structural Equation Modeling*, International Series in Operations Research & Management Science 267, https://doi.org/10.1007/978-3-319-71691-6_7

relevant; and the predictive orientation of PLS-SEM offers the potential to answer practical questions. Based on three highly ranked published articles in management accounting, I demonstrate how PLS-SEM can be useful in this field. First, Ittner et al. (1997) illustrate how accounting research can use archival data in PLS-SEM; second, Hartmann and Maas (2011) construct an example of the exploratory use of PLS-SEM in management accounting; and third, Burkert and Lueg (2013) use hierarchical latent construct in PLS-SEM in their management accounting research.

7.1 Introduction

Management accounting research addresses the provision of financial and non-financial information to managers for decision-making (Chapman et al. 2007). In contrast to financial accounting, management accounting produces reports that comprise cost analyses and financial forecasts for the organization's internal recipients. This information is often very detailed. There is a strong relationship between financial and management accounting because a company's internal steering is closely connected with external financial accounting, such as value-based management (VBM). In its purest form, the aim of VBM is for all processes in the internal steering of a company to serve shareholder value maximization (Burkert and Lueg 2013). These illustrated areas of management accounting research indicate that the field is very heterogeneous.

Nevertheless, theoretical knowledge is often scarce in management accounting research (Smith and Langfield-Smith 2004). There is no specific management accounting theory such as a theory of organizational incentive systems that can explain how to design and use an incentive system (Malmi and Granlund 2009). Thus, management accounting researchers are forced to rely on theories from other research fields, which they must adjust by including different management accounting-relevant constructs (cf. Chenhall and Smith 2011). This also means that management accounting researchers are frequently confronted by uncertainty about both the correctness of the research model and the correctness of the construct measurements (Malmi and Granlund 2009; Luft and Shields 2014). It follows that research models in management accounting must test many different possible relationships (Shields 2015; Luft and Shields 2003, 2014). For example, the functionality of management control systems can be influenced by many different factors, such as different types of control mechanisms (Otley 2016; Speklé 2001). However, in contrast to the scarce theoretical knowledge, the basis for the data is habitually very rich in management accounting research, as it is in financial accounting (Moers 2007).

This situation, in which the basis for the data is rich and the theoretical knowledge is scarce, was one of the main forces behind the development of PLS-SEM (Henseler et al. 2014). Lohmöller and Wold (1980) describe the exploratory approach as an evolutionary process in which the approach extracts new knowledge from the data and puts flesh on a theoretical skeleton. This type of exploratory model development is characterized by the introduction of a new construct, an item, or a new path connection or by the exclusion of one of them (Wold 1985; Jöreskog 1993). Because PLS-SEM is less restrictive than covariance-based SEM (CB-SEM), an explorative strategy is much easier to follow (Rigdon 2013; Hair et al. 2017a). Therefore, Smith and Langfield-Smith (2004) argue that PLS-SEM appears to be tailor-made for the exploratory state in management accounting research. For example, for the abovementioned theory of organizational incentive systems, it would be possible to test different possible research models (path connections, constructs, items) with the aim of finding a comprehensive set of propositions for explaining incentive systems under different contexts (Malmi and Granlund 2009).

Smith and Langfield-Smith (2004) find only one contribution using PLS-SEM for data analysis in top-tier accounting journals up to 2001. In contrast, Nitzl (2016) finds 37 contributions using PLS-SEM in top-tier accounting journals up to 2013, which demonstrates the accelerated use of PLS-SEM in management accounting research in recent years. Although the use of PLS-SEM is accelerating in management accounting research, it is not yet prevalent. There appear to be different reasons for this reluctance to use PLS-SEM in management accounting. Thus, the aim of this contribution is both to discuss the past use of PLS-SEM in management accounting research and to illustrate promising avenues for using it in the future. To understand the limited use of PLS-SEM in management accounting research, possible reasons for this reluctance must also be discussed. There remains a strong misunderstanding of what PLS-SEM delivers that is different from covariance-based structural equation modelling (CB-SEM) -biased estimators (Smith 2015). Therefore, PLS-SEM is viewed as an inferior method (Richter et al. 2016). Important reasons for using PLS-SEM, such as the ability to test complex models and its causal-predictive orientation, are seldom recognized. However, these reasons are very useful for answering research questions in management accounting.

Although there are already contributions in management accounting journals that can provide an orientation for the fruitful use of PLS-SEM in future research, Ittner et al. (1997) was the first management accounting study that used PLS-SEM. This study provides an example of how accounting research can use PLS-SEM on archival data. The authors identify the factors that determine the relative weights given to financial and non-financial measures in managerial compensation. The aim of the study accords with the characteristic of PLS-SEM as a method that maximizes the explained variance of endogenous constructs. Hartmann and Maas (2011) and Burkert and Lueg (2013) are examples of how PLS-SEM can be used in an exploratory way. Hartmann and Maas (2011) use a step-wise approach to show how PLS-SEM can be modified to deliver deeper insights through a more complex model. Burkert and Lueg (2013) had to develop a new multi-dimensional construct measurement for use in value-based management (VBM), which is in line with the explorative characteristic of PLS-SEM (Sarstedt et al. 2016). Furthermore, the authors examine the influence of CFOs on VBM.

After an overview of the use of PLS-SEM in management accounting provided in Sect. 7.2, in Sect. 7.3, I discuss possible reasons behind the reluctance to use

PLS-SEM in management accounting research. In Sect. 7.4, the high potential for the use of PLS-SEM in management accounting will be illustrated using the three abovementioned examples. Finally, Sect. 7.5 concludes this contribution.

7.2 The Use of PLS-SEM in Management Accounting in Existing Literature

Nitzl (2016) reviews 37 articles that had been published up-to 2013 in 11 top-tier management accounting journals. In accordance with Hair et al. (2012), he analyses these articles based on the following criteria: the reasons for using PLS-SEM, the characteristics of the data analysed, the characteristics of the model used, the model evaluation, and reporting matters.

Regarding the rationale for using PLS-SEM in management accounting, the most commonly mentioned reasons in the articles reviewed are small sample sizes and the non-normal distribution of data, which indicates a very technical view of the usefulness of PLS-SEM. Whereas PLS-SEM shows better power for estimations for small sample sizes than CB-SEM (Reinartz et al. 2009), reasons connected to the non-normal distribution of data are no longer valid because robust algorithms have been generated for CB-SEM in the case of non-normally distributed data (Goodhue et al. 2012). However, other important reasons for using PLS-SEM are rarely discussed, such as the use of formative measurements, the ability to estimate more complex models, and the prediction orientation. However, these more rarely mentioned reasons are central to why research should use PLS-SEM (Henseler et al. 2014, 2016a). As a composite approach, PLS is preferable either when formative measurements, or newly developed constructs are on hand (Sarstedt et al. 2016; Richter et al. 2016), or when prediction is also an important element of answering a research question (Shmueli et al. 2016).

In accordance with the rationale for using PLS-SEM for small sample sizes, the average sample size in PLS-SEM is 138, which is considerably smaller than the 292 for CB-SEM (Henri 2007). Important data characteristics are often not in focus. For example, only a small number of studies provide a power test to indicate that the PLS-SEM has an adequate sample size. However, an adequate sample size is important for theory development in any research field (Willaby et al. 2015). Hence, although PLS-SEM analyses can deliver initial results beginning with a sample size of 20, a determination should also be made regarding whether the PLS-SEM has a sufficient sample size for detecting effects. Nitzl (2016) shows that more than 40% of his reviewed studies do not have the critical sample size that would allow them to detect relevant effects. Another issue that involves analysing data from a survey study with the help of PLS-SEM is control for common measure variance (CMV). Only a few contributions address this topic in management accounting research (for example, Mahama and Cheng 2013). Whereas various methods frequently fail to control for CMV, one promising method is to use a marker variable in a PLS-SEM research model (Chin et al. 2013).

In comparison with marketing, PLS-SEM in management accounting shows a lower use of latent variables but a higher number of inner path relationships (Hair et al. 2012). This observation reflects the exploratory state of management accounting, in which theoretical development is often weak (Shields 2015). Another aspect is that formative measurements are less used in management accounting research, which is somewhat surprising because formative measurements capture characteristics that should be of special interest in this field, such as the measurement of performance (Nitzl and Chin 2017). An extremely large number of studies in management accounting show the potential misspecification of measurements (Rodgers and Guiral 2011). In addition, management accounting research must develop more reliable and valid measurements for future theory development in the field (Bisbe et al. 2007).

In terms of model evaluation, management accounting researchers have typically used traditional criteria to evaluate measures in the past. However, there are new criteria for evaluating measures, such as the heterotrait-monotrait ratio of correlation (HTMT) (Henseler et al. 2015). HTMT is much more reliable for evaluating the discriminant validity of measurements in PLS-SEM than the oft-used Fornell-Larcker criteria (Fornell and Larcker 1981). For evaluating PLS-SEM, management accounting research has relied mainly on the coefficient of determination (R^2), which remains an important criterion. However, in the future, this statistic might be supplemented with newly developed criteria, for example, goodness-of-fit statistics such as the standardized root mean square residual (SRMR) (Henseler et al. 2014). Furthermore, analyses of moderating and mediating effects are of central importance to the evaluation of models. Only a small number of the reviewed articles include accurate analyses of such effects. Currently, there are important updated guidelines for analysing moderation and mediation effects (Nitzl et al. 2016; Fassott et al. 2016).

Finally, reporting plays a central role in PLS-SEM in management accounting research. Reporting the crucial criteria is essential for the reproduction of results and theory development (Shields 2015). Nearly all studies in management accounting provide information about the sample structure, the model structure, and the measurements used. However, in addition to this important information, information about the computational choices should be given (Chin 2010). Such information can include the weighting scheme used or the bootstrap method employed. This information is necessary to replicate a study's findings.

7.3 Reasons for the Reluctance to Use PLS-SEM in Management Accounting Research

There are several reasons for the reluctance of accounting scholars to use PLS-SEM. The first is that accounting research, like financial accounting, is dominated by the use of regression analysis (Oler et al. 2010; Lee et al. 2011). Regression analysis has proven to be a useful tool for analysing data (Smith 2015). However,

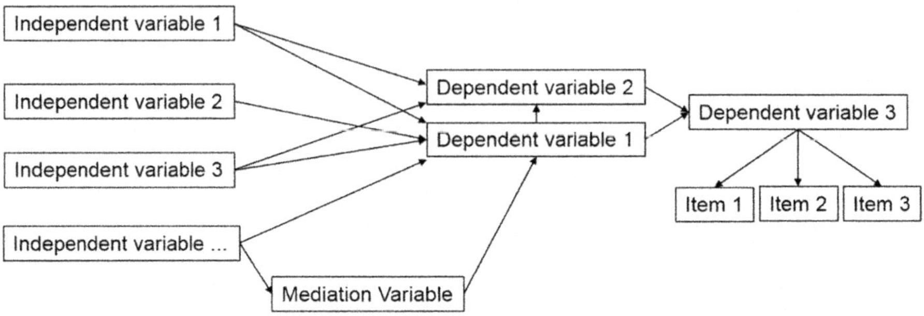

Fig. 7.1 A more sophisticated research model

there are several limitations in the application of regression analysis that have prevented the development of more sophisticated research models. Regression analysis is primarily used to test the relationship between a singular dependent variable and several independent variables. The aim of a regression model is both to explain as much as possible about the dependent variable and to determine how important certain independent variables are in this explanation as indicated by correlation coefficients.

For the development of future theory, however, it is necessary to develop a more holistic picture of causes and effects (Luft and Shields 2014). For more than one dependent variable, such as different performance measures, separate regression equations must be built. This limits the possibility of theory testing and development (Smith and Langfield-Smith 2004). As research models in management accounting have become more sophisticated – such as when more than one dependent variable needs to be analysed, when latent variables need to be included in the research model, or when mediation processes should be analysed—it is often not possible to use regression. This type of research model is illustrated in Fig. 7.1.

The dominance of regression analysis creates the risk that management accounting researchers will make inferior choices and systematically ignore important theoretical relationships that are critical for developing a more holistic picture (Chenhall 2012). However, in PLS-SEM it is quite easy to test such a model, as illustrated in Fig. 7.1. Through its flexibility, PLS-SEM allows us to explore more complex relationships and therefore to open new possibilities for further theoretical development in management accounting.

A further reason for accounting scholars' reluctance to use PLS-SEM can be found in the critical views regarding survey studies (Guffey and Harp 2017; Van der Stede et al. 2005). The use of structural equation modelling is tightly connected with such survey studies (Smith and Langfield-Smith 2004). Survey-based data are often viewed as subjective, whereas archival-based data are often seen as objective (Moers 2007). This is true despite the fact that survey-based research often provides the only possibility to obtain important insights into organizational structures and internal performance management and how the two are connected to external reporting.

These types of data can often be collected only through surveys. Additionally, archival and secondary data seldom fulfil the requirements of a factor-based method such as CB-SEM (Churchill 1979; Richter et al. 2016). Hence, it appears that management accounting researchers often do not use SEM to test their research models (Hampton 2015). In contrast to CB-SEM, PLS-SEM allows researchers to integrate data from financial statements, such as assets, expenses, or revenues, with internal data from documents collected by the company in the context of doing business, such as internal key performance indicators (KPIs) (Rigdon 2013; Goh et al. 2014; Moers 2007), thus making it much more interesting for use in management accounting research.

Other reasons for the reluctance to use PLS-SEM in accounting seem to be a misunderstanding of certain characteristics of PLS-SEM and the misuse of PLS-SEM in the past. In accounting, PLS-SEM is often considered a "poor man's" SEM (Smith 2015) because it produces biased estimators in the presence of reflective measurements when it is treated as a common factor measurement. However, reflective measurements such as trust, intention to use or satisfaction are only partially relevant in management accounting research. For other measurements, such as formative measurements or newly developed reflective measurements, PLS-SEM as a composite approach is a better choice (Sarstedt et al. 2016). Furthermore, whenever advanced software applications are made easily available, there is the danger that statistical methods will be misused (Marcoulides and Chin 2013). PLS-SEM offers a high degree of flexibility in terms of answering research questions with fewer restrictions. Hence, there is inevitably a greater risk that the method will be misused. Researchers should be aware that restrictive statistical assumptions might give other researchers the wrong impression in terms of finding a theoretically correct model. Nonetheless, even when a research model shows a good global fit index (GFI), for example, there may be alternative models with substantially different explanations but the same or even better GFI values (Chin 1998).

Table 7.1 compares PLS-SEM, CB-SEM, and regression analysis in light of the current characteristics of management accounting research.

As shown in Table 7.1, the characteristics of PLS-SEM match the current situation in management accounting research like a hand in a glove. However, some words of caution are also necessary. A small group of researchers are very critical about the use of PLS-SEM (e.g., McIntosh et al. 2014; Rönkko and Evermann 2013; Rönkkö et al. 2015, 2016). These proponents of CB-SEM assume that under a common factor model, CB-SEM is superior because it not only accounts for measurement errors for each indicator but also models correlated errors (McIntosh et al. 2014). Moreover, critics argue that CB-SEM often fails to converge at a solution in the case of model misspecifications, whereas PLS seldom has such convergence problems. Hence, they argue that PLS may lead to an increased acceptance of bad models. Therefore, a researcher who follows this argumentation of a common factor perspective should be very careful about using PLS-SEM (Richter et al. 2016). In line with Rigdon (2016), the misconception about PLS-SEM stems mainly from a particular view of an estimation philosophy.

Table 7.1 Comparison of different methods with regard to the characteristics of management accounting research

Characteristics of management accounting research	PLS-SEM	CB-SEM	Regression analysis
Management accounting research must test complex models because the theoretical bases are often weak, and they frequently must use newly developed construct measurements	PLS-SEM is appropriate both for research situations in which theoretical knowledge is scarce and for newly developed construct measurements	CB-SEM is appropriate for research situations in which theoretical knowledge is rich, and well-developed structural and construct measurements must be tested	Regression analysis allows the integration of many independent variables
Important constructs such as performance are often formative by nature	As a composite approach, PLS-SEM easily allows the modelling of formative measurements	As a common factor approach, CB-SEM's use of formative measurements is very limited	Under regression analyses, it is not possible to use multi-item construct measurements such as formative measurements
Archival data play an important role in management accounting research	The integration of archival data is easily possible	Archival data do not regularly follow the assumption of common factors	Archival data can be easily integrated
The sample size in management accounting, especially in survey-based studies, is often low (<200)	PLS-SEM shows better power to detect effects for small sample sizes (e.g., 100) and has already provided initial results from very small sample sizes (e.g., 20)	CB-SEM requires larger sample sizes (e.g., 250 or higher)	Like PLS-SEM, regression analyses can also be used with small sample sizes (e.g., 100)
The aim in management accounting as a practically oriented research field is to provide the best possible explanation of dependent variables	PLS-SEM aims to maximize the explanation of one or more dependent variables	CB-SEM aims to minimize the discrepancy between the covariance matrix implied by the sample data and the covariance matrix implied by a theoretical model	Regression analyses aim to maximize the explanation of a single dependent variable

One additional consideration regarding the reluctance to use PLS-SEM that has not been discussed so far is that most management accounting researchers are not yet familiar with it. The reviewed contribution by Nitzl (2016) indicates that the authors who have used PLS-SEM are highly interconnected. Thus, there seems to be only a small group of management accounting researchers who are trained in and have used PLS-SEM in their contributions. Significant changes in the understanding and use of PLS-SEM are ongoing. In particular, there are several new textbooks that explain step-by-step how researchers can use PLS-SEM (e.g. Hair et al. 2017b). Hence, it seems to be only a matter of time before a larger group of researchers will become more aware of the useful characteristics of PLS-SEM and will use it in management accounting research.

7.4 Practical Examples of the Use of PLS-SEM in Management Accounting

7.4.1 Using Archival Data in PLS-SEM

The use of PLS-SEM in management accounting is typically connected to survey studies (Nitzl 2016). Notably, Ittner et al. (1997) is the first study in a top-tier management accounting journal that used PLS-SEM. It is the only study to use archival data in a broader sense. This is astounding because unlike CB-SEM, it is also possible to use archival data in PLS-SEM, as mentioned in Table 7.1. Archival data play a central role in management accounting research and are frequently understood as more valid than survey data (Moers 2007). Based on the connection to survey studies and the relevance of archival studies in management accounting research, the use of SEM seems to be low in contrast to other research fields (Henri 2007; Moers 2007).

Ittner et al. (1997) examine the influencing factors in CEO bonus contracts that determine the relative weights given to financial and non-financial measures. This research question is of special interest because bonus contracts are typically based on financial metrics, such as revenue or earnings per share. However, non-financial measures such as customer or employee satisfaction have become increasingly relevant in recent years. Furthermore, a sole focus on financial metrics has come under fire recently because of its (over-) emphasis on short-term results. Financial metrics may also not contain enough information, such as the motivations for managers to align their actions with the owner's aims. Therefore, for firms, it could be interesting to incorporate non-financial metrics in bonus contracts because such measures can inform owners about managerial efforts beyond financial performance. Another reason to examine non-financial metrics might be that powerful CEOs incorporate them to obtain a higher bonus than would be legitimized by the company's financial situation.

Ittner et al. (1997) test the influence of six different constructs on the weight that is given to financial and non-financial metrics. These factors include (1) the organizational business strategy, (2) the quality strategy, (3) regulations, (4) financial performance, (5) exogenous noise from performance measures, and (6) CEO influence. As discussed above, the most interesting component given the background of a PLS-SEM analysis is the use of archival data in construct measurement by Ittner et al. (1997). Each SEM analysis must start with such a measurement of constructs (Hair et al. 2017b). The typical approach to constructing measures is a common factor (reflective measurement), which is defined by the shared variance of items. Such common factors are frequently used for measuring psychological constructs, such as trust or attitudes (Hampton 2015; Nitzl and Hirsch 2016). Another approach that contrasts with the common factor method is to build composites of items. Thus, a composite factor is represented as the weighted sum of its items (Chin 1995). PLS-SEM is based on composites. If the underlying true population is following a common factor, then the results of PLS-SEM will be biased (Dijkstra 2010). However, in cases in which the measures are in a more exploratory stage, then composite measurement PLS-SEM is a more accurate method (Sarstedt et al. 2016).

For studies using archival data, the assumptions for common factor measurement are rarely fulfilled, and therefore such data are frequently not appropriate for common factor measurements (Churchill 1979; Rigdon 2013). Thus, for analysing archival data, PLS-SEM is an important alternative. Archival data can be used to build informative composite measurements. Such construct measurements are typically man-made constructs in which the construct definition depends on a variety of items that are included in the measurement. One common example of such measurements in management accounting is performance measurements (Nitzl and Chin 2017). In performance measurements, the questions concern the relevant factors for success or determining valid proxies for measurement, which means that all potential items for measuring performance may not be included in the composite measurement. However, the measure should include those factors that are generally viewed as the most relevant for a certain business or that can best serve as proxies (e.g., Speklé and Verbeeten 2014).

Ittner et al. (1997) use the following four items to measure an organization's competitive strategy: (1) the ratio of research and development costs to sales, (2) the market-to-book ratio, (3) the ratio of the number of employees to sales, and (4) the number of product or service introductions. These various items illustrate different aspects of measuring a competitive strategy. Higher values for each of the items mean a greater orientation towards the prospective strategy. Furthermore, the composition of these items can be viewed as a proxy because other different items can also be relevant for measuring a business strategy. This approach is completely in line with the explanation provided above regarding composite measurements, and they should be modelled in an SEM as a formative measurement, although Ittner et al. (1997) specify the measurement of organizational strategy as a reflective measurement and thus use a common factor approach. Because the basis of the measurement is formative by nature, the critical values of the loadings are below the critical value of 0.7 in that study.

Despite the incorrect specification of the construct measurements in Ittner et al. (1997), the study remains a prototype for using PLS-SEM in combination with archival data. Using archival data matches perfectly with using PLS-SEM as a composite approach in which different sources of data can be easily integrated: data from financial statements, from document analyses or from organizations that have gathered for the purpose for doing business. Including such information in an SEM makes it possible to answer research questions that extend beyond the domain of a simple regression analysis; this ability is of special interest when multiple dependent variables are used, when different items are used to build a construct, or when a mediating analysis should be performed.

7.4.2 *The Exploratory Use of PLS-SEM*

An important argument for using PLS-SEM is its exploratory characteristics (see Table 7.1). These characteristics were one of the main reasons for the development of PLS-SEM, in that it can be used to extract new knowledge from data when other methods fail (Lohmöller and Wold 1980). However, the exploratory characteristics of PLS-SEM should not be confused with an exploratory factor analysis (EFA), for example, in which common factors are produced based on certain criteria. EFA is often recommended for testing when constructs differ from one another in connection with an SEM. However, it is not advisable to mix results from an EFA with a PLS-SEM analysis due to model over-fitting problems and the capitalization of chance. An exploratory approach in the sense of an EFA must always be cross-validated with new data to confirm the findings (Marcoulides and Chin 2013). In contrast to an EFA, the exploratory approach of a PLS-SEM is characterized by (a) its relatively high flexibility in testing complex path models because it is less constrained by identification or other technical issues than other methods, (b) its robustness to misspecificationv and therefore more appropriate for newly developed construct measurements, and (c) its greater power for small sample sizes (Henseler et al. 2014; Reinartz et al. 2009).

An example of how PLS-SEM can be used in an exploratory manner can be found in Hartmann and Maas (2011). These authors examine how uncertainty influences the use of budgets and the role of a controller in a company. First, the authors develop three hypotheses based on theoretical assumptions, which also reflects the confirmatory basis of a PLS-SEM. It is a prerequisite of PLS-SEM that a causal relation must be formulated based on theory or conceptual frameworks. Without a theoretical basis, the analysis and data collection are blind (Zimmerman 2001). Theories provide guidance regarding which variables should be considered in the model and causal relationships. Hence, it is not advisable to begin with a model in which each variable is connected to every other variable because of the possibility of different causal ordering (Ittner 2014).

After building the basic research model as a skeleton with three constructs and three hypotheses, Hartmann and Maas (2011) delve deeper into the details. In particular, they divide each construct into further sub-constructs. For example, the construct "role of budgeting" is divided into "enabling budget use" and "coercive budget use". This construct offers more detailed insights and has more practical relevance. The basic relationships are based on their theoretical derivations, but when every sub-construct is connected to each of the other sub-constructs, the theoretical basis is often weak, and it cannot be used to obtain a detailed perspective on the possible relationships (Shields 2015). This approach by Hartmann and Maas (2011) matches perfectly with the exploratory characteristics of PLS-SEM and the aim of developing a more holistic view and a tool to develop theory in management accounting (Luft and Shields 2014).

Under their exploratory approach, Hartmann and Maas (2011) obtain many interesting findings, including how the different sub-constructs of uncertainty influence the role of budgeting. However, one problem with such an exploratory approach is that effects beyond direct effects can occur. The possibility that certain constructs influence subsequent constructs indirectly via other constructs increases with model complexity. Increased model complexity is extremely common in exploratory studies. Such effects are called mediating effects. If a study is more exploratory in nature, in an SEM, more attention can be paid to such indirect effects (Nitzl et al. 2016). Hartmann and Maas (2011) argue that uncertainty has no influence on the role of the controller as a policeman, but that is true only when one is considering direct effects. For the effect via enabling budgeting use, there is a mediating effect. Hence, uncertainty also impacts the controller's role as policeman.

Another argument for the use of PLS-SEM in Hartmann and Maas (2011) is the use of newly developed construct measurements. The authors develop new construct measurements for "enabling budget use" and "coercive budget use", which is a common practice for a research area that is in an exploratory state. However, for newly developed construct measurements, the potential for misspecification is higher (Rodgers and Guiral 2011). The development of constructs requires multiple rounds and a sound theoretical basis (Bagozzi 2011). PLS-SEM, as a limited information procedure, is more robust to misspecification (Gerbing and Hamilton 1994). Hence, it is also advisable to use PLS-SEM when common factor measurements, such as reflective measurements, constitute the parts of a model that have not yet been widely tested (Sarstedt et al. 2016).

Finally, another topic that is frequently connected to exploratory management accounting research is small sample sizes, particularly in survey studies (Van der Stede et al. 2005). A main reason that PLS-SEM is employed is that it can deliver results even for small sample sizes (Chin and Newsted 1999). However, to detect a true effect, a sufficient sample size is critical (Willaby et al. 2015) because the larger the sample size, the higher the probability of detecting effects. To determine the necessary sample size, a researcher can revert to a power analysis in PLS-SEM (Nitzl 2016). Based on its complexity, the required sample size is 85 for the research model of Hartmann and Maas (2011). Because the authors use a sample size of 134 to test their model, it is unlikely that they miss relevant effects.

7.4.3 *Using PLS-SEM for Hierarchical Latent Constructs*

A hierarchical latent construct is characterized in the simplest form by the fact that a certain number of first-order constructs are used to measure a second-order construct. Both the first-order and the second-order constructs can be defined as reflective or formative measurements. However, only certain combinations appear to be relevant for modelling in a structural equation model (Lee and Cadogan 2013). The most relevant combination in PLS-SEM seems to measure the first-order constructs as reflective measurements and the second-order constructs as formative measurements (Becker et al. 2012; Ringle et al. 2012). Using hierarchical latent constructs allows for greater theoretical parsimony and reduces model complexity (Wetzels et al. 2009). In management accounting research, the use of hierarchical latent constructs will play an important role in future theory development (Nitzl 2016). Their use further allows matching across different levels of abstraction in a model. However, there is also a theoretical danger of using hierarchical latent variables when they are not embedded in a nomological network (Chin 1998).

Burkert and Lueg (2013) offer a good illustration of using hierarchical latent constructs in management accounting research. The authors develop a hierarchical latent construct to measure the degree of sophistication in the VBM in a company. VBM is defined as a system in which decisions are based on the objective of shareholder value maximization. VBM comprises all relevant processes and pertinent information systems and links them to the aim of shareholder value maximization by means of setting targets and formulating action plans (Ittner and Larcker 2001). Although companies may implement the same system, the use of the system can vary among companies. Measuring different dimensions of VBM aims to overcome the problem of simply modelling the introduction of a new system in terms of "adoption versus non-adoption" (Ansari et al. 2010).

Based on a literature review, Burkert and Lueg (2013) develop a preliminary sample of first-order constructs to measure VBM. This approach is in line with the argument in Table 7.1 in favour of using PLS-SEM for a newly developed construct. These preliminary constructs are pretested with the help of different groups of practitioners and academics. Finally, the first-order constructs comprise the following six constructs: (1) Portfolio: the selection of a strategy among alternative strategies based on the highest expected value added to shareholder value; (2) F_Valuedr: the provision of information on pertinent generic financial value drivers; (3) NF_Valuedr: the provision of information on relevant company-specific non-financial value drivers or key performance indicators; (4) Actionplan: empowerment and the development of action plans based on key performance indicators; (5) Targetset: target setting that is focused on long-term value creation including synergies; and (6) Mindset: the establishment of a value-based mindset among all employees. These first-order constructs are measured using items based on Likert scales.

Although the contribution by Burkert and Lueg (2013) is an innovative approach to measuring relevant topics in management accounting research, there are some problematic technical topics in modelling the hierarchical latent variable using

PLS-SEM in their contribution. First, the authors assume a reflective measurement of the second-order VBM construct using the six first constructs. This type of modelling for hierarchical latent constructs is understood to be problematic in the literature (Lee and Cadogan 2013). Reflective measurements should thus be interchangeable and act as common measures of the underlying construct. This approach conflicts with the view that first-order constructs are distinct dimensions. As discussed above, formative measurements are characterized by the fact that they comprise different elements of the phenomenon under consideration. For example, it is possible in a company that provisions are made for financial drivers (factor 2) but not for non-financial drivers (factor 3). Hence, the first-order constructs reflect different aspects by nature and should be used as formative measurements of the second-order VBM construct.

Based on a relatively high Cronbach's alpha and high composite reliability, Burkert and Lueg (2013) assume that the six first-order constructs define a reflective measurement of the second-order construct VBM. This conclusion must be viewed as problematic. The authors do not consider that the high values for Cronbach's alpha and composite reliability might have their origins in a common measured variance (CMV). CMV is an important topic when a single person answers the items for the measures of the six first-order constructs. With a single respondent, there is a propensity to answer in a certain manner regardless of the type of dimension. Therefore, high correlations between the six first-order constructs may not indicate a reflective measurement. The type of measurement in hierarchical latent variables between the dimensions should be based on theoretical considerations (Chin 1998).

To model the second-order VBM construct, Burkert and Lueg (2013) use the averages of the six multi-item measurements of the first-order construct because each construct in a PLS-SEM must be measured by items, in contrast to CB-SEM. This workaround must be viewed as problematic because when using the average value of the items—which is used for one second-order construct—the information regarding the error is no longer within the statistical consideration for measuring the second-order construct. Therefore, a better workaround for modelling hierarchical second-order constructs is the repeated indicator use approach (Lohmöller 1989). Becker et al. (2012) show that the repeated use of indicators is the most valid approach in a context in which the first-order constructs are measured reflectively and the second-order constructs are measured formatively using the first-order constructs. This finding means that the indicators of the six reflective first-order constructs in this example are used as a block, and once again, as formative measurements of the second-order VBM construct. This analysis must be undertaken in a separate PLS-SEM model when the second-order construct is an endogenous variable in the research model. Otherwise, the coefficient of determination is 1 and it is not possible for another antecedent construct in the model to explain anything regarding the second-order construct. Furthermore, all first-order constructs should be measured with the same number of items. Otherwise, the first-order constructs with more items will have a higher weight in the second-order construct.

7.5 Conclusions

As shown in this contribution, in recent years, PLS-SEM has become an important tool in management accounting research, despite the fact that management accounting remains dominated by other tools such as regression analysis, which is often used to analyse archival data. Additionally, PLS-SEM is often viewed as an inferior method to CB-SEM in the field of management accounting. It is argued that PLS-SEM delivers biased estimators. Against this backdrop, the possibilities of using PLS-SEM to answer research questions in management accounting remain underestimated.

PLS-SEM can also be used to analyse archival data. In contrast to regression analysis, however, archival data can be connected through more complex relationships using PLS-SEM. The criticism that PLS-SEM delivers biased estimators holds only when common factors serve as the benchmark. As shown in this contribution, many of the construct measurements used in management accounting must be understood as composite measurements. Such construct measurements (e.g., performance measures) play a central role in management accounting research. For such construct measurements, PLS-SEM is more appropriate than CB-SEM.

After discussing these possible reasons for management accounting scholars' reluctance to use PLS-SEM, three examples of fruitful ways to use PLS-SEM in management accounting were provided. Our discussion of the drawbacks of the cited contributions illustrates some of the dangers related to the future use of PLS-SEM. PLS-SEM has developed rapidly in recent years, and these prior contributions could not have considered these new developments. For example, important improvements include the new HTMT criteria for evaluating discriminant validity (Henseler et al. 2015), the development of a GFI (Henseler et al. 2016a), tests of model invariance for group tests (Henseler et al. 2016b), new criteria for using PLS-SEM for predictions (Shmueli et al. 2016), or the use of PLS-SEM for analysing longitudinal data (Roemer 2016). In addition, there are new guidelines for analysing moderating and mediating effects in PLS-SEM (Fassott et al. 2016; Nitzl et al. 2016). These newly elaborated findings should be considered in future management accounting studies.

Against the backdrop of this expansion of PLS-SEM, there are still certain drawbacks to PLS-SEM compared to regression analysis. Regression analysis offers additional tools for analysing data that have not been regularly considered in PLS-SEM research. Working with financial data in management accounting, for example, requires additional tests for the absence of heteroscedasticity, or in the case of time series data, a Durbin-Watson d-statistic (Smith 2015). However, many methods used in regression analyses to examine archival data should be transferable without major problems.

Finally, management accounting researchers should recognize that there are additional interesting tools for PLS-SEM, such as the importance-performance matrix analysis (IPMA). IPMA can be used to illustrate the relationship between the impact and value of one variable compared to a dependent variable (Ringle and Sarstedt 2016). This tool gives managerial accounting researchers the opportunity both to achieve a deeper understanding of the data and to better communicate their results. Other interesting tools in PLS-SEM include approaches to detecting unobserved heterogeneity such as finite-mixture PLS (FIMIX-PLS) (Sarstedt et al. 2011) and prediction-oriented segmentation (PLS-POS) (Becker et al. 2013). Addressing heterogeneity is important to ensure that the results are not biased (e.g., by different groups) (Schlittgen et al. 2016). The usefulness of these additional tools has not yet been considered in top management accounting journals.

References

Ansari, S. M., Fiss, P. C., & Zajac, E. J. (2010). Made to fit: How practices vary as they diffuse. *Academy of Management Review, 35*(1), 67–92.

Bagozzi, R. P. (2011). Measurement and meaning in information systems and organizational research: Methodological and philosophical foundations. *MIS Quarterly, 35*(2), 261–292.

Becker, J.-M., Klein, K., & Wetzels, M. (2012). Hierarchical latent variable models in PLS-SEM: Guidelines for using reflective-formative type models. *Long Range Planning, 45*(5–6), 359–394.

Becker, J.-M., Rai, A., Ringle, C. M., & Völckner, F. (2013). Discovering unobserved heterogeneity in structural equation models to avert validity threats. *MIS Quarterly, 37*(3), 665–694.

Bisbe, J., Batista-Foguet, J.-M., & Chenhall, R. (2007). Defining management accounting constructs: A methodological note on the risks of conceptual misspecification. *Accounting, Organizations and Society, 32*(7–8), 789–820.

Burkert, M., & Lueg, R. (2013). Differences in the sophistication of value-based management: The role of top executives. *Management Accounting Research, 24*(1), 3–22.

Chapman, C. S., Hopwood, A. G., & Shields, M. D. (2007). *Handbook of management accounting research* (Vol. 1, p. 560). Oxford: Elsevier.

Chenhall, R. H. (2012). Developing an organizational perspective to management accounting. *Journal of Management Accounting Research, 24*(1), 65–76.

Chenhall, R. H., & Smith, D. (2011). A review of Australian management accounting research: 1980–2009. *Accounting and Finance, 51*(1), 173–206.

Chin, W. W. (1995). Partial least squares is to LISREL as principal components analysis is to common factor analysis. *Technology Studies, 2*(2), 315–319.

Chin, W. W. (1998). Commentary: Issues and opinion on structural equation modeling. *MIS Quarterly, 22*(1), xii–xvi.

Chin, W. W. (2010). How to write up and report PLS analyses. In V. Esposito Vinzi, W. W. Chin, J. Henseler, & H. Wang (Eds.), *Handbook of partial least squares: Concepts, methods and applications* (pp. 655–690). Heidelberg: Springer.

Chin, W. W., & Newsted, P. R. (1999). Structural equation modeling analysis with small samples using partial least squares. In R. H. Hoyle (Ed.), *Statistical strategies for small sample research* (pp. 307–341). Thousand Oaks: Sage.

Chin, W. W., Thatcher, J. B., Wright, R. T., & Steel, D. (2013). Controlling for common method variance in PLS analysis: The measured latent marker variable approach. In H. Abdi, W. W. Chin, V. Esposito Vinzi, G. Russolillo, & L. Trinchera (Eds.), *New perspectives in partial least squares and related methods* (pp. 231–239). Heidelberg: Springer.

Churchill, G. A. (1979). A paradigm for developing better measures of marketing constructs. *Journal of Marketing Research, 16*(1), 64–73.

Dijkstra, T. K. (2010). Latent variables and indices: Herman wold's basic design and partial least squares. In V. Esposito Vinzi, W. W. Chin, J. Henseler, & H. Wang (Eds.), *Handbook of partial least squares: Concepts, methods and applications: Vol. 2. Springer handbooks of computational statistics series* (pp. 23–46). Heidelberg: Springer.

Fassott, G., Henseler, J., & Coelho, P. S. (2016). Testing moderating effects in PLS path models with composite variables. *Industrial Management & Data Systems, 116*(9), 1887–1900.

Fornell, C. G., & Larcker, D. F. (1981). Structural equation models with unobservable variables and measurement error: Algebra and statistics. *Journal of Marketing Research, 18*(3), 328–388.

Gerbing, D. W., & Hamilton, J. G. (1994). The surprising viability of a simple alternate estimation procedure for construction of large-scale structural equation measurement models. *Structural Equation Modeling: A Multidisciplinary Journal, 1*(2), 103–115.

Goh, C. F., Ali, M. B., & Rasli, A. (2014). The use of partial least squares path modeling in causal inference for archival financial accounting research. *Jurnal Teknologi, 68*(3), 57–62.

Goodhue, D. L., Lewis, W., & Thompson, R. (2012). Does PLS have advantages for small sample size or non-normal data? *MIS Quarterly, 36*(3), 891–1001.

Guffey, D. M., & Harp, N. L. (2017). The journal of management accounting research: A content and citation analysis of the first 25 years. *Journal of Management Accounting Research.* In press. doi:https://doi.org/10.2308/jmar-51592.

Hair, J. F., Sarstedt, M., Ringle, C. M., & Mena, J. A. (2012). An assessment of the use of partial least squares structural equation modeling in marketing research. *Journal of the Academy of Marketing Science, 40*(3), 414–433.

Hair, J., Hollingsworth, C. L., Randolph, A. B., & Chong, A. (2017a). An updated and expanded assessment of PLS-SEM in information systems research. *Industrial Management & Data Systems, 117*(3), 442–458.

Hair, J. F., Hult, G. T. M., Ringle, C. M., & Sarstedt, M. (2017b). *A primer on partial least squares structural equation modeling (PLS-SEM)* (2nd ed.). Thousand Oaks: Sage.

Hampton, C. (2015). Estimating and reporting structural equation models with behavioral accounting data. *Behavioral Research in Accounting, 27*(2), 1–34.

Hartmann, F. G. H., & Maas, V. S. (2011). The effects of uncertainty on the roles of controllers and budgets: An exploratory study. *Accounting and Business Research, 41*(5), 439–458.

Henri, J.-F. (2007). A quantitative assessment of the reporting of structural equation modeling information: The case of management accounting research. *Journal of Accounting Literature, 26*, 76–115.

Henseler, J., Dijkstra, T. K., Sarstedt, M., Ringle, C. M., Diamantopoulos, A., Straub, D. W., Ketchen, D. J., Hair, J. F., Hult, G. T. M., & Calantone, R. J. (2014). Common beliefs and reality about PLS: Comments on Rönkkö & Evermann (2013). *Organizational Research Methods, 17*(2), 182–209.

Henseler, J., Ringle, C. M., & Sarstedt, M. (2015). A new criterion for assessing discriminant validity in variance-based structural equation modeling. *Journal of the Academy of Marketing Science, 43*(1), 115–135.

Henseler, J., Hubona, G. S., & Pauline, A. R. (2016a). Using PLS path modeling in new technology research: Updated guidelines. *Industrial Management & Data Systems, 116*(1), 2–20.

Henseler, J., Ringle, C. M., & Sarstedt, M. (2016b). Testing measurement invariance of composites using partial least squares. *International Marketing Review, 33*(3), 405–431.

Ittner, C. D. (2014). Strengthening causal inferences in positivst field studies. *Accounting, Organizations and Society, 39*(7), 545–549.

Ittner, C. D., & Larcker, D. F. (2001). Assessing empirical research in managerial accounting: A value-based management perspective. *Journal of Accounting and Economics, 32*(1), 349–410.

Ittner, C. D., Larcker, D. F., & Rajan, M. V. (1997). The choice of performance measures in annual bonus contracts. *The Accounting Review, 72*(2), 231–255.

Jöreskog, K. G. (1993). Testing structural equation models. In K. A. Bollen & J. S. Long (Eds.), *Testing structural equation models* (pp. 294–316). Newbury Park: Sage.

Lee, N., & Cadogan, J. W. (2013). Problems with formative and higher-order reflective variables. *Journal of Business Research, 66*(2), 242–247.

Lee, L., Petter, S., Fayard, D., & Robinson, S. (2011). On the use of partial least squares path modeling in accounting research. *International Journal of Accounting Information Systems, 12*(4), 305–328.

Lohmöller, J.-B. (1989). *Latent variable path modeling with partial least squares*. Heidelberg: Physica.

Lohmöller, J-B., & Wold, H. (1980). *Three-mode path models with latent variables and partial least squares (PLS) parameter estimation*. Paper presented at the European Meeting of the Psychometric Society, Groningen, Netherlands, June 18–21, 1980.

Luft, J., & Shields, M. D. (2003). Mapping management accounting: Graphics and guidelines for theory-consistent empirical research. *Accounting, Organizations and Society, 28*(2), 169–249.

Luft, J., & Shields, M. D. (2014). Subjectivity in developing and validating causal explanations in positivist accounting research. *Accounting, Organizations and Society, 39*(7), 550–558.

Mahama, H., & Cheng, M. M. (2013). The effect of managers' enabling perceptions on costing system use, psychological empowerment, and task performance. *Behavioral Research in Accounting, 25*(1), 89–114.

Malmi, T., & Granlund, M. (2009). In search of management accounting theory. *European Accounting Review, 18*(3), 597–620.

Marcoulides, G. A., & Chin, W. W. (2013). You write, but others read: Common methodological misunderstandings in PLS and related methods. In H. Abdi, W. W. Chin, V. Esposito Vinzi, G. Russolillo, & L. Trinchera (Eds.), *New perspectives in partial least squares and related methods* (pp. 31–64). Heidelberg: Springer.

McIntosh, C. N., Edwards, J. R., & Antonakis, J. (2014). Reflection on partial least squares path modeling. *Organizational Research Methods, 17*(2), 210–251.

Moers, F. (2007). Doing archival research in management accounting. In C. S. Chapman, A. G. Hopwood, & M. D. Shields (Eds.), *Handbooks of management accounting research* (Vol. 1, pp. 399–413). Oxford: Elsevier.

Nitzl, C. (2016). The use of partial least squares structural equation modelling (PLS-SEM) in management accounting research: Directions for future theory development. *Journal of Accounting Literature, 39*, 19–35.

Nitzl, C., & Chin, W. W. (2017). The case of partial least squares (PLS) path modeling in managerial accounting research: Back to the future. *Journal of Management Control, 28*(2), 137–156.

Nitzl, C., & Hirsch, B. (2016). The drivers of a superior's trust formation in his subordinate: The manager-management accountant example. *Journal of Accounting & Organizational Change, 12*(4), 472–503.

Nitzl, C., Roldán, J. L., & Cepeda, G. (2016). Mediation analyses in partial least squares structural equation modeling: Helping researchers to discuss more sophisticated models. *Industrial Management & Data Systems, 116*(9), 1849–1864.

Oler, D. K., Oler, M. J., & Skousen, C. J. (2010). Characterizing accounting research. *Accounting Horizons, 24*(4), 635–670.

Otley, D. (2016). The contingency theory of management accounting and control: 1980–2014. *Management Accounting Research, 31*, 45–62.

Reinartz, W. J., Haenlein, M., & Henseler, J. (2009). An empirical comparison of the efficacy of covariance-based and variance-based SEM. *International Journal of Research in Marketing, 26*(4), 332–344.

Richter, N. F., Cepeda, G., Roldán, J. L., & Ringle, C. M. (2016). European management research using partial least squares structural equation modeling (PLS-SEM). *European Management Journal, 34*(6), 589–597.

Rigdon, E. E. (2013). Partial least squares path modeling. In G. R. Hancock & R. O. Mueller (Eds.), *Structural equation modeling: A second course* (2nd ed., pp. 81–116). Charlotte: Information Age.

Rigdon, E. E. (2016). Choosing PLS path modeling as analytical method in European management research: A realist perspective. *European Management Journal, 34*(6), 598–605.

Ringle, C. M., & Sarstedt, M. (2016). Gain more insight from your PLS-SEM results: The importance-performance map analysis. *Industrial Management & Data Systems, 116*(9), 1865–1886.

Ringle, C. M., Sarstedt, M., & Straub, D. W. (2012). A critical look at the use of PLS-SEM in MIS quarterly. *MIS Quarterly, 36*(1), iii–xiv.

Rodgers, W., & Guiral, A. (2011). Potential model misspecification bias: Formative indicators enhancing theory for accounting researchers. *The International Journal of Accounting, 46*(1), 25–50.

Roemer, E. (2016). A tutorial on the use of PLS path modeling in longitudinal studies. *Industrial Management & Data Systems, 116*(9), 1901–1921.

Rönkko, M., & Evermann, J. (2013). A critical examination of common beliefs about partial least squares path modeling. *Organizational Research Methods, 16*(3), 425–448.

Rönkkö, M., McIntosh, C. N., & Antonakis, J. (2015). On the adoption of partial least squares in psychological research: Caveat emptor. *Personality and Individual Differences, 87*, 76–84.

Rönkkö, M., McIntosh, C. N., Antonakis, J., & Edwards, J. R. (2016). Partial least squares path modeling: Time for some serious second thoughts. *Journal of Operations Management, 47–48*, 9–27.

Sarstedt, M., Becker, J.-M., Ringle, C. M., & Schwaiger, M. (2011). Uncovering and treating unobserved heterogeneity with FIMIX-PLS: Which model selection criterion provides an appropriate number of segments? *Schmalenbach Business Review, 63*(1), 34–62.

Sarstedt, M., Hair, J. F., Ringle, C. M., Thiele, K. O., & Gudergan, S. P. (2016). Estimation issues with PLS and CBSEM: Where the bias lies! *Journal of Business Research, 69*(10), 3998–4010.

Schlittgen, R., Ringle, C. M., Sarstedt, M., & Becker, J.-M. (2016). Segmentation of PLS path models by iterative reweighted regressions. *Journal of Business Research, 69*(10), 4583–4592.

Shields, M. D. (2015). Established management accounting knowledge. *Journal of Management Accounting Research, 27*(1), 123–132.

Shmueli, G., Ray, S., Velasquez Estrada, J. M., & Chatla, S. B. (2016). The elephant in the room: Predictive performance of PLS models. *Journal of Business Research, 69*(10), 4552–4564.

Smith, M. (2015). *Research methods in accounting* (3rd ed.). London: SAGE.

Smith, D., & Langfield-Smith, K. (2004). Structural equation modeling in management accounting research: Critical analysis and opportunities. *Journal of Accounting Literature, 23*, 49–86.

Speklé, R. F. (2001). Explaining management control structure variety: A transaction cost economics perspective. *Accounting, Organizations and Society, 26*(4), 419–441.

Speklé, R. F., & Verbeeten, F. H. M. (2014). The use of performance measurement systems in the public sector: Effects on performance. *Management Accounting Research, 25*(2), 131–146.

Van der Stede, W. A., Young, S. M., & Chen, C. X. (2005). Assessing the quality of evidence in empirical management accounting research: The case of survey studies. *Accounting, Organizations and Society, 30*(7–8), 655–684.

Wetzels, M., Odekerken-Schroder, G., & van Oppen, C. (2009). Using PLS path modeling for assessing hierarchical construct models: Guidelines and empirical illustration. *MIS Quarterly, 33*(1), 177–195.

Willaby, H. W., Costa, D. S., Burns, B. D., MacCann, C., & Roberts, R. D. (2015). Testing complex models with small sample sizes: A historical overview and empirical demonstration of what partial least squares (PLS) can offer differential psychology. *Personality and Individual Differences, 84*, 73–78.

Wold, H. (1985). Partial least squares. In S. Kotz & N. L. Johnson (Eds.), *Encyclopedia of statistical sciences* (Vol. 6, pp. 581–591). New York: Wiley.

Zimmerman, J. L. (2001). Conjectures regarding empirical managerial accounting research. *Journal of Accounting and Economics, 32*(1), 411–427.

Index

N.K. Avkiran, C.M. Ringle (eds.), *Partial Least Squares Structural Equation Modeling*, International Series in Operations Research & Management Science 267, https://doi.org/10.1007/978-3-319-71691-6

Printed by Printforce, the Netherlands